W9-DAY-833

THE
SIMON AND SCHUSTER
QUESTION
AND ANSWER BOOK

THE
SIMON AND SCHUSTER
QUESTION
AND ANSWER BOOK

By Kathleen N. Daly

Wanderer Books
Published by Simon and Schuster, New York

Published by WANDERER BOOKS
A Simon & Schuster Division of
Gulf & Western Corporation
Simon & Schuster Building
1230 Avenue of the Americas
New York, New York 10020

Designed by Stanley S. Drate

Manufactured in the United States of America

10 9 8 7 6 5 4 3 2 1

WANDERER and colophon are trademarks
of Simon & Schuster

Library of Congress Cataloging in Publication Data

Daly, Kathleen, N.
    The Simon and Schuster question and answer book.

    Includes index.
    Summary: A compilation of facts in question and answer format on such topics as science and
technology, outer space, prehistoric times, animal behavior, and the supernatural.
    1.Science—Juvenile literature. 2. Children's questions and answers. [1. Science. 2. Questions and
answers] I. Title.
Q163.D25   500   82-4830
ISBN 0-671-44427-1   AACR2

CONTENTS

KEEP CHECKING!

The world is so full of remarkable things that a whole book of questions and answers could be written about an ordinary day in your life. Your body is miraculous. The clock and the radio are remarkable. The breakfast pop-up toaster is a marvel. Central heating and/or air conditioning, electric light, and elevators would seem like miracles to your ancestors of not so long ago. Modern transport, such as the school bus, the family car, airplanes overhead and subways below ground, are fantastic. And how about the telephone and television that receive radio waves from man-made satellites in orbit far above the earth?

There's no need to look far for marvels and wonders—provided you have an inquiring mind.

You'll be amazed at how much more you will get out of your reading when you keep a "question-and-answer" approach in your mind. Seeking the right question to ask about a subject is a wonderful way of concentrating the mind!

You'll find the words, "Keep checking!" after many answers in this book. That's because new things are happening every day, and facts and figures change overnight. For example, not so long ago earth was calculated to be 2 billion years old, but more recent dating methods now put its age between 3.5 and 4.5 billion years. In 1980 a team of scientists discovered some fossilized organisms that were estimated to be 3.5 billion years old, so it is very likely that earth itself is a much older planet than we thought.

Whenever you read a book or look up a reference and something strikes you as interesting, puzzling, or astounding, first find out exactly when the book was published. (The date is usually given with the copyright notice on the back of the title page.) Then, if you can, try to find a more up-to-date book on the subject that interests you. Check with your parents, teachers, or librarians for the most recent facts. Much information may also be found at museums, zoos, and other places where knowledgeable people are usually happy to answer your questions. Get into the habit of reading a reliable daily newspaper and weekly or monthly magazines that report on new discoveries. Such periodicals are available in the public library.

But don't be surprised if the facts and figures are slightly different in every source you check! It's almost impossible to be 100 percent accurate in dealing with, for example, the staggering figures that are used to describe dimensions in outer space, or the age of ancient fossils or rocks. People do the best they can, often making an intelligent guess based on comparing facts and figures from different sources.

As this book goes to press, space probes *Voyager 1* and *Voyager 2* are on their way to the far-distant planets Uranus, Neptune, and possibly Pluto. Meanwhile, scientists are still analyzing the thousands of photographs and other data sent back by probes several years ago. Almost every day new facts come to light!

Back on earth, the search still goes on for the definitive ancestor of modern man. There is great controversy about this subject and great rivalry between teams of fossil hunters.

Another factor to bear in mind is that no matter how reliable your source, there actually may be several different answers to any question, each one accurate in a special way.

For example, what is the tallest building on earth? Is it the Sears Tower in Chicago—or is it the tiny hut built by a Sherpa mountaineer near the top of Mount Everest, the highest mountain on earth? Or is it a structure such as a radio tower? And does the height of the building include an antenna on its roof?

The fact is, people often say or write things that are intended to be sensational, rather than strictly true. Sensational "facts" will capture your interest, of course, and that's fine—provided you then become curious to know more and thus seek out other sources of information. That's why I keep repeating throughout this book, "Keep checking!"

OUTER SPACE

How hot is the sun? Is it the hottest spot in the solar system? (No! There is one place at least three hundred times hotter!) Are there canals on Mars? Are there diamonds in space?

Thanks to spacecraft carrying complicated instruments, new photographic processes, electronic computers, radio telescopes, and other marvels, we are constantly finding out more about the universe. But with each new discovery, there are more questions to be asked and answered.

Keep checking!

Q *What star seems biggest to the naked eye?*

A The sun. What we call the sun is a star, and a fairly small one compared to many millions of others in the galaxy. Some of the other stars are thousands of times larger than the sun, but they are so far away they look like pinpricks in the sky.

Q *In what galaxy does our solar system exist?*

A The Milky Way Galaxy. If you look up at the sky on a clear evening, you can see what appears to be a densely speckled band of stars arching across the sky. Long ago, people who saw this arch compared it to a stream of milk poured across the sky, and the name, Milky Way, has stuck. What you are really seeing is the side view of our galaxy, looking edge-on to its disc.

Q *Are the sun and the moon the same size?*

A No. The sun is about four hundred times bigger than the moon. Its diameter is 865,000 miles (1,392,000 kilometers), while the moon's diameter is 2,160 miles (3,476 kilometers). The sun, however, is four hundred times farther away from us than the moon, so it appears to be about the same size as a full moon in the sky. [Caution: never try to look at the sun with your naked eyes—or even through sunglasses! Even at a distance of about 92,950,000 miles (150 million kilometers), the sun can burn the delicate tissues of your eyes, causing blindness.]

Q *What is a light year?*

A A light year is the distance light travels during one year. Light travels at a speed of 186,282 miles (300,000 kilometers) per second. That equals about 6 million million miles (9.46 million million kilometers) per year. Astronomers measure distances in light years because the normal units (miles, feet, kilometers, etc.) are much too small for recording the vast distances of space.

Q *How long does it take for light from the sun to reach earth?*

A From six to eight minutes. But think about this: the Milky Way, the galaxy to which our solar system belongs, is so huge that it takes 100,000 *years* for light to travel from one end to the other!

Q *How hot is the sun?*

A The surface temperature of the sun is about 6,000°C (10,000°F). At the core it is probably at least 14,000,000° C (25,000,000°F).

Q *Is the sun burning?*

A No, not in the way we think of a fire burning. Deep inside the sun, nuclear reactions are taking place that cause energy to radiate from it and keep it glowing. Although the sun is halfway through its lifetime, it still has enough fuel to keep it shining for another 5,000 million years.

Q *What are sunspots and how do they affect earth's radio and television services?*

A Sunspots are regions on the sun's surface that are cooler than the surrounding areas. The spots occur where strong magnetic forces from within the sun break through to the surface. When sunspots occur in large numbers, they trigger eruptions. The eruptions form clouds of gases. These explosions send out streams of tiny particles that reach hundreds of millions of miles to earth and beyond.

The particles interfere with the light and radio waves that reach our radio and television sets. They may also cause rainfall shortages and temperature changes. And sometimes they cause streaks of color in the night sky at the northern and southern poles of the world. These lights are called auroras.

Q *What is the nearest star to the earth, apart from the sun?*

A Alpha Centauri. It is 25 million million miles (40 million million kilometers) or 4.3 light years away.

Q *How long would it take for people on earth to send a "light" message to Alpha Centauri and receive an answer?*

A The answer is: more than 8.5 years—from, "Hello, Alpha Centauri!" to, "Hello, Planet Earth!" (That's 4.3 light years multiplied by 2.)

Q *What is the difference between a star and a planet?*

A A planet is a solid, non-luminous sphere revolving around a star and shining by light reflected from the star or stars.

In our solar system, nine planets orbit our star, the sun.

The word *planet* comes from the Greek word for "wanderer."

Planets are called wanderers because we can see their movements in the night sky.

Stars are so far away that they appear to be fixed in exact clusters which we call constellations.

A star is a hot, burning sphere of gas. The central core of a star burns with tremendous energy at temperatures of millions of degrees. A burning star sends forth light and other forms of nuclear energy, such as electromagnetic waves.

Our star, the sun, is only one of many trillions of stars in the universe.

Q *What is the difference between a moon and a planet?*

A A planet revolves around a sun; a moon revolves around a planet. As far as we know, all planets are smaller than their suns, and all moons are smaller than their planets.

Q *How many stars are in the Milky Way Galaxy?*

A About a trillion. Astronomers come to this conclusion by counting the visible stars in a small section of the galaxy and then increasing the number in proportion to the complete size of the galaxy as a whole.

Q *What is a solar eclipse?*

A A solar eclipse happens when the moon moves directly between the earth and the sun. The moon blocks out light from the sun for a short time. The moon's shadow falls on the earth. As the moon moves in its orbit, the shadow moves across the earth. The shadow is small, so only a small part of the earth lies in its shadow. It takes only a few minutes for the solar eclipse to be over.

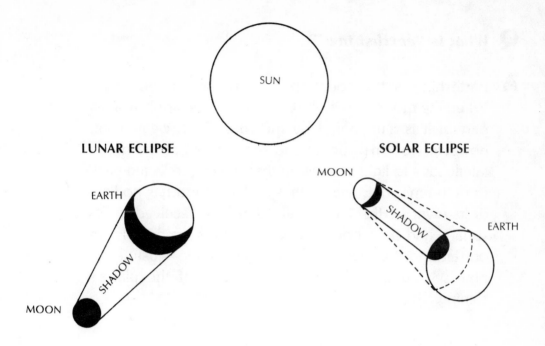

LUNAR ECLIPSE

SUN

SOLAR ECLIPSE

EARTH

MOON

SHADOW

MOON

SHADOW

EARTH

Q *What is a lunar eclipse?*

A A lunar eclipse happens when the earth moves between the sun and the moon. The earth's shadow stops all sunlight from reaching the moon. Lunar eclipses are visible only at night. The earth's shadow starts at one edge of the moon. The entire moon is soon covered by the shadow. A lunar eclipse lasts several hours, because the earth's shadow is bigger than the diameter of the moon. A lunar eclipse may be seen by many people from many places on earth.

Q *What is "earthshine"?*

A Earthshine is the counterpart of moonlight on earth. When the moon is in its thin, crescent phase, only a small part of it is illuminated by the sun. On a clear night, however, you may be able to see the rest of the moon faintly, due to light from earth that is being reflected back from the moon. Since earth is a planet, and has no light of its own, earthshine is really sunshine received on its surface and then bounced back up to the moon. From out among the planets, Earth would shine—just as we now see Jupiter, Venus, Mars, or any of the planets shining.

Q *Do all the planets have satellite moons?*

A As far as we know, Mercury and Venus have no moons. Mars has two, Jupiter has sixteen, Saturn has at least twenty-one—maybe twenty-three, Uranus has five, Neptune has two, and Pluto, we think, has one. When *Voyager 2* sends back reports from Uranus and Neptune, perhaps many more planetary satellites will be counted.

Q *What do we mean by a "space probe"?*

A A space probe is an unmanned spacecraft rocketed into space and equipped with hundreds of different kinds of instruments that send back information to earth. Most space probes make "flybys"—that is, the craft does not actually land on a surface. Different instruments on the craft are sensitive to infrared, visible, and ultraviolet light, cosmic rays, and solar winds.

Space probes that do land on a planet (or other surface) have "arms" and "hands" that can pick up soil and rocks. Instruments then send back information about their weight and composition.

Space-probe cameras and other instruments do not transmit images that you and I would understand. They send complicated signals that computers on earth can then "translate" into detailed single or multiple photographs on screens that look like television screens. The computers are able to convert the different wavelengths into color.

The data received takes months or even years for scientists and their computers to analyze. But at last the distant lights of the sky are being reformed by incredible machines into pictures of hard rocks, ice, whirling gases, erupting volcanoes, unexpected moons, circling rings, and other marvels of the cosmos.

Q *Which moon is the largest in the solar system?*

A Ganymede, one of Jupiter's moons. It has a diameter of 3,120 miles (5,276 kilometers). It was discovered in 1610 by the astronomer Galileo. Saturn's moon, Titan, with a diameter of 2,980 miles (5,120 kilometers), is a close second, while Earth's moon, with a diameter of 2,160 miles (3,476 kilometers), is small by comparison, though of course it looks large to us because it is relatively close.

Q *Which was the first planet to be explored by a space-probe flyby?*

A Venus is the most brilliant of our planets, and, before the Space Age, was the most mysterious. A thick cover of eternal cloud makes it impossible to see the surface of Venus clearly.

In 1962 *Mariner 2*, launched by the United States, flew by Venus at a distance of about 21,000 miles (34,000 kilometers). After that there were numerous Soviet and U.S. probes.

In 1975 two Soviet craft, *Venera 9* and *Venera 10*, became the first craft ever to land on the surface of Venus. Miraculously, they survived long enough to send back some data and one photograph each.

In March 1982 *Venera 13* added eight more pictures of the surface of Venus, some through multiple filters that showed color, including brownish rocks covered with fine dust and a bluish surface.

Q *What have the space probes told us about Venus?*

A Mostly that it is extremely hot (about 900° F or 500° C average surface temperature). That's hot enough to melt lead!

The atmosphere of Venus is so dense that almost no sunlight reaches its surface, and so heavy that a human being would be crushed like a balloon, if not first melted by the heat.

The surface of Venus is relatively flat, but it also has craters, canyons, and a mountain range as high as Mount Everest on Earth.

There is no hope of life, as we know it, on this roasting planet.

Q *Could there be life, as we know it, on Mars?*

A No. Our first photographs of Mars were sent back by several U.S. *Mariner* probes, notably *Mariner 9,* in 1971. Over seven thousand pictures received are still being analyzed.

In 1976, U.S. *Vikings 1* and *2* made the first successful landings on Mars.

Photographs show that Mars is a barren, rocky desert, with great valleys, terraced polar icecaps, and towering volcanoes. The planet's violent dust storms can be seen from telescopes on Earth.

Soil and rock samples analyzed by the probes show no signs of life.

Q *Are there artificial canals on Mars?*

A No. For many years astronomers wondered if the lines perceived by telescopes on the surface of Mars could have been built by extraterrestrial beings. Space-probe photographs, however, disclose that these "canals" were gouged out by natural forces such as wind and water.

Q *Is there water on Mars—and if so, could it support certain forms of life?*

A Whatever water there is on Mars is locked up on the polar caps and perhaps in a layer of permafrost beneath the surface. Life forms dependent on water would have to be able to tap these frozen supplies. But, as far as we know, there is no life on Mars.

Q *What is the hottest spot ever discovered in the solar system?*

A A cloud of electrified gases circling Saturn. Its temperature is three hundred times hotter than the sun's outer regions. The discovery was reported by scientists analyzing data sent back to Earth by *Voyager 2* in 1981. Temperatures ranged from 600 million to more than 1 billion degrees Fahrenheit! Scientists cannot yet explain the reason for this sizzling new surprise from space.

Q *What is the only moon in the solar system known to have an atmosphere?*

A Saturn's huge moon, Titan. In available photographs Titan seems to be covered by a kind of cosmic smog. The smoggy atmosphere is probably made up of nitrogen and methane. It blankets a surface that could have oceans of liquid methane.

Q *Of the solar system's nine planets, which ones have not been examined by space probes?*

A Uranus, Neptune, and Pluto. These are the three planets farthest from the sun and from the earth. They were discovered in relatively modern times, being almost impossible to see with the naked eye. Uranus was discovered in 1781, Neptune in 1846, and Pluto, a real newcomer, in 1930.

Scientists hope to learn more about Uranus and Neptune when *Voyagers* 1 and 2 reach them in the late 1980's.

Pluto, 39.44 million miles (7,375 million kilometers) from the sun, is the smallest, coldest planet. Its diameter is about half that of Earth.

THE SOLAR SYSTEM

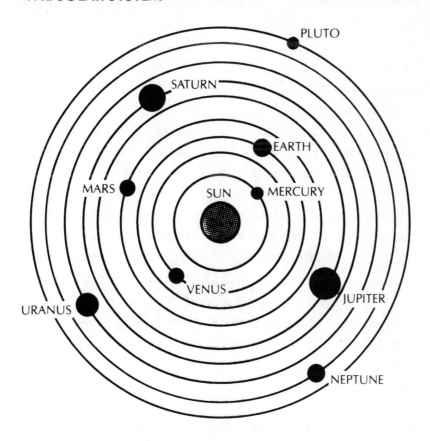

On an average, it takes Pluto 248 *years* to revolve around the sun! (Mercury, the planet nearest to the sun, takes only 88 *days* to revolve around it.) Earth takes 365¼ days.

Q *Where in space is "The Great Red Spot"?*

A On Jupiter, the largest and most massive of the planets. Its diameter at the equator is 142,800 kilometers compared to Earth's 6,787 kilometers. Its mass is 318, compared to Earth's 1.

In 1979, U.S. space probes *Voyager 1* and *2* showed that the Great Red Spot is a permanently whirling storm. Its surface area is greater than that of Earth!

Jupiter is probably the most colorful planet, with clouds of gases in shades of orange and pink, constant lightning, and a thin ring of dust circling the planet.

On one of Jupiter's major moons, Io, scientists were surprised to see violent volcanic activity, with plumes of smoke and flame rising thousands of miles into space.

Scientists think that fine particles from Io's volcanoes may account, in part, for the faint ring of dust around Jupiter.

Q *What did the Voyagers discover about Saturn, the second largest planet?*

A Pictures from space showed that the famous rings of Saturn are made up of enormously complex lines, braids, grooves like those on a record, and huge gaps. The rings are formed by ice particles orbiting the planet.

Q *How long will it take* Voyager 2 *(the satellite that explored Saturn and Jupiter in spectacular flybys in 1980 and 1981) to reach Uranus and Neptune?*

A If it continues to keep to its remarkable schedule (it was only 3.1 seconds behind schedule after traveling four years across 1.24 billion miles of space), *Voyager 2* will reach Uranus in 1986 and Neptune in 1989.

Q *Are there diamonds in space?*

A Quite possibly! Scientists think that both Uranus and Neptune, giant planets four times the size of Earth, are about one-fifth carbon, the stuff of which diamonds are made. Pressures could vary from 200,000 times to 6 million times the pressure of Earth's atmosphere.

The diamonds, if they exist, could be in the form of glittering flakes drifting through the dense lower atmosphere of the two planets. Or the diamonds could have fallen to the surface of the planets, encrusting the rocky core.

It's not likely, however, that these diamonds will be harvested any time soon!

Q *What is an asteroid?*

A An asteroid is a very small planet. More than two thousand asteroids have been tracked. In the solar system, the largest one orbiting the sun is Ceres: it has a diameter of 470 miles (750 kilometers).

An asteroid is different from a planet in that it may have an irregular shape; that is, it may be much less spherical than most of the planets.

There may be millions more asteroids that we have yet to learn about.

Q *What is a comet?*

A A comet is somewhat like a large, dirty snowball hurtling in orbit around the sun. Its body, or nucleus, is made of ice, frozen gases, and dust. When it gets near the sun, its frozen gases burn up, and solar winds form the vaporized gases into a tail. Some comets and their tails, millions and millions of miles long, are spectacular sights.

The word *comet* means long-haired star, and that's just what a comet looks like: a bright star with long "hair," that is, vaporized gases, streaming behind it.

Q *What is the most famous comet?*

A Halley's comet. Its first appearance near earth was first recorded in 240 B.C. Every 76 years it appears in our skies on its trip around the sun.

Halley's comet is due to make its next appearance in 1986, and it may be seen without the aid of a telescope.

Q *When was the first known and recorded instance of a comet that seemed to be colliding with the sun?*

A The presumed collision occurred in August 1979. The event was seen and photographed by an experimental instrument called Solwind. Solwind studies all the sun's activity.

The comet passed through the instrument's field of vision, streaking toward the sun at a speed of about 640,000 miles (1,000,000 kilometers) per hour, with its tail millions of miles long flaming behind it.

Scientists say that the comet was probably one of the "sun-grazers," a group of comets whose orbits often skim by the sun. Some gravitational quirk, perhaps a tug from an asteroid, may have nudged the comet into what is thought to be its fatal crash-course with the sun.

Q *What is a "falling star" or a "shooting star"?*

A It is a meteoroid—a small chunk of matter falling through space. When a meteoroid enters the earth's atmosphere, it encounters friction and becomes very hot and glowing. We see it as a streak of light in the sky.

Q *When we see a meteoroid streaking by, is the flash happening now, or did it occur light years ago?*

A The light that we see is within the earth's atmosphere, so its cause and effect happened only a fraction of a second ago.

Q *What happens to a meteoroid when it hits the earth?*

A Very few meteoroids ever reach earth. They are falling so fast and burning with such tremendous heat that they burn up into tiny dust particles. About *25 million* meteoroids fall into earth's atmosphere every day!

The few meteoroids that survive the trip are called meteorites.

About 50,000 years ago a huge meteorite fell near Winslow, Arizona. It formed a crater almost a mile (1.6 kilometers) in diameter, and about 600 feet (183 meters) deep.

Fortunately, such events occur infrequently.

Q *Why do people in the northern hemisphere see meteor showers mainly in August?*

A Because in August the earth passes through the orbit of a comet named P/Swift-Tuttle. Meteors are debris (bits) blasted off comets as they orbit the sun. Of the many comets that scatter their debris into earth's atmosphere, the so-called Perseids of P/Swift-Tuttle seem to be the most reliable in presenting their annual meteor shows.

Q *If space is so full of dust, meteors, comets, and asteroids, why has no man-made space vehicle been hit by them?*

A Space vehicles have been hit by dust particles, but the particles made only tiny craters, called microcraters, that can be seen only under a microscope.

It is possible that the impact of a small particle on *Voyager 2* may have interrupted data transmission for a few days as the craft flew by Saturn in 1981.

On the whole, we could say that space is so vast and man-made spacecraft are so tiny that the chances of a collision with other than dust particles are very slim.

Q *How can a rocket escape from earth's gravitational pull?*

A By exerting a force *greater* than the pull of earth's gravity. Earth's gravitational force extends millions of miles out into space, so a rocket must not only reach *escape velocity* [a speed of about 25,000 miles (40,000 kilometers) per hour], but maintain this speed, or it will fall back down to earth.

Q *What is the difference between an astronaut and a cosmonaut?*

A Very little. Astronaut means "sailor of the stars." Cosmonaut means "sailor of the cosmos," or universe.

U.S. explorers of space are called astronauts. Soviet explorers are called cosmonauts. These people know how to fly their craft, how to survive under almost any circumstances, how to operate hundreds of machines, how to cook, sew, perform surgery, how to land and take off from various surfaces, and how to find their way back to earth or to a space laboratory.

Q *What was so important about astronauts landing on the moon?*

A It was the first time that men had landed on an extraterrestrial surface.

Almost as important, the moon-landing project was one of the best planned projects in history. Thousands of scientists spent many years preparing for the historic event. Their efforts were successful, and millions of people all over the world were able to see it happen on their television sets. All were gripped with excitement and awe.

This was not science fiction—it was science *fact!*

Q *When did the first moon-landing take place?*

A On July 20, 1969.

Q *Who were the astronauts involved in the moon-landing?*

A Neil Armstrong was the first man to set foot on the moon from the lunar module *Eagle*. He was followed by his partner, Edwin Aldrin. Michael Collins was the astronaut who orbited the moon in the mother ship awaiting the others' return.

Their spacecraft was called *Apollo 11*.

Q *What were the first words uttered by a man on the moon?*

A "That's one small step for a man, one giant leap for mankind," said Neil Armstrong, as he placed his foot on the lunar surface. (Since there are no winds on the moon, his footprints, and those of Edwin Aldrin, and the imprint of their landing craft, will remain undisturbed indefinitely.)

Q *Why must astronauts on the moon talk to each other by radio, even though they are only a few feet apart?*

A Because sound waves can travel only through air, water, metal, or anything that they can bounce upon. They travel by transmitting energy from molecule to molecule (and even air is made up of molecules) and by sending out ripples, the way a pebble hitting water sends out ripples or waves. Sound waves cause the air, metal, eardrum, or whatever, to vibrate back and forth.

   In space, and on the surface of the moon, there is no air, so sound waves have nothing to bounce against.

Q *Is it really possible to send and receive radio messages from outer space?*

A Yes. Invisible waves, such as ultraviolet waves, gamma rays, and radio waves are being sent to us constantly from outer space. Planet Earth is protected from the more harmful rays by a layer of atmosphere, but rays still "leak" to and from it.

For example, radio waves from earth began when "radio" was invented, about 1920. Since then, radio and television waves have been beaming out into the atmosphere—and beyond. Some extraterrestrial intelligence may have picked up the waves of your favorite TV show!

Q *Is it possible that extraterrestrial beings have already visited earth?*

A Yes. It is possible. But so far, scientists cannot find any real proof of ancient visitations, in spite of so-called landing strips, pictures of unidentified flying objects (UFOs), and strange-looking creatures depicted in ancient artworks.

Q *Is it possible that one day people may be able to form settlements in space?*

A Yes. We have already sent up numerous satellites that receive and transmit communications to and from earth. Experimental space stations, such as the Skylabs, have been lived in by astronauts for several months.

Men landed on the moon (in 1969), walked on it, and returned to earth safely.

Space probes have traveled farther and farther out into space, sending back valuable information about our solar system.

Space shuttles can travel into space and return to earth, to be used again. They may be the airplanes and buses of the future.

Q *What would it be like to live on the moon?*

A You'd have to carry your own atmosphere with you all the time!

People would have to live in airtight structures that could withstand extraordinary changes of temperature—from 212° F (100° C) in the day to −247° F (−155° C) at night. (And remember that one cycle of day to night on the moon lasts about 28 earth days!)

To hop from one building to another, you'd have to wear a spacesuit.

Plants would have to grow in *hydroponic* greenhouses, without soil. Water in these growing areas would contain all the nutrients that are found in the soil on earth. The plants would help make oxygen for people to breathe.

All waste would be carefully recycled for future use. Everybody would have to exercise to keep muscles in trim—with a reduced gravity, people would quickly lose the tone of the strong muscles that we need to keep the body functioning actively on earth.

Q *What is a "black hole"?*

A A black hole is a star at the end of its life cycle: it has collapsed in upon itself. All of its enormous mass is now concentrated into a volume much smaller than the original size of the star.

When matter is squeezed together into a small volume, it becomes incredibly dense. Even a tiny piece would have tremendous weight. When a star has collapsed into a "black-hole" state, it is so dense that its gravitational pull stops even light from leaving its surface!

Q *If a black hole is invisible, how do we know it exists?*

A By scientific observation and calculation. Astronomers first figured out that black holes existed because they observed galaxies and stars in orbit around "something," but there appeared to be nothing there for the objects to orbit around—unlike the earth and other planets orbiting around the shining sun in our solar system. Since they couldn't see that "something," they called it a black hole. The existence of black holes is still something of a mystery to scientists.

Q *If there is life on other planets or in other galaxies, would the creature have two eyes and a nose and mouth and walk on legs the way people do?*

A There is no reason to think so. There are thousands of different forms of life on earth, besides the two-legged variety that is *Homo sapiens,* our own kind, or species.

Just for example: there are ocean creatures that can live under more than a thousand tons of pressure, in utter darkness, and in near-freezing temperatures. (One tiny creature, *Neopililina galathenea,* wasn't discovered until 1951.)

There are plants, such as Venus's-flytrap and the pitcher plants, that trap and eat insects.

The African lungfish can live for more than four years without water!

Some creatures don't need mates of the opposite sex in order to reproduce their own species!

And think of the amazing life cycles of butterflies, moths, and other insects, who change from caterpillars to winged creatures!

With all this remarkable variety of life forms on earth, it is almost impossible for us to guess what life may be like in other galaxies.

It would certainly be very conceited of us to think that life forms should have the same physical features as *Homo sapiens!*

PLANET EARTH

Among the millions of stars and planets in our galaxy there is one small planet, Earth, that glows with color in the vast coldness of space. Photographs from spacecraft show us the blue of the oceans, the white of the shifting cloud patterns, the greens and browns of the continents.

Earth is the only place where "earthlings" can live—unless they carry their own atmosphere with them in capsules or spacesuits.

What makes our planet so different from all the other planets?

Find out the answer to that question and others about thunder and lightning, icebergs, earthquakes, volcanoes, ocean tides, coral reefs, and other wonders of the world in which we live.

Q *What makes Planet Earth different from any other planet in the solar system?*

A Planet Earth is different in at least three major factors. One is the atmosphere, a layer of gases that surrounds our planet. The earth's atmosphere protects us from dangerous cosmic rays. The gases of which the atmosphere is composed are breathed by all plants and animals.

Another extraordinary thing about our planet is its vast amount of water: 75 percent of the globe is covered with water.

And then, of course, there is the fact that there are millions of life forms here, from microscopic bacteria to mighty whales, from insects and fishes to birds and people, from tiny mosses to giant trees.

Refer to the chapter on outer space to find out how very different the other planets are!

Q *What does earth look like from outer space?*

A It looks like a glowing blue-and-white globe. We didn't really know this until the first man-made satellites were shot up above the atmosphere and sent back pictures of the planet we call home. Later, astronauts (from the U.S.) and cosmonauts (from the U.S.S.R.) described in their own words the wonder of earth glowing with color in the cold emptiness of space.

Q *What is the "greenhouse effect"?*

A Scientists have described the gradual heating of earth's atmosphere as the "greenhouse effect." The heating is caused by carbon dioxide in the atmosphere, given off by the burning of fossil fuels such as coal.

The layer of carbon dioxide acts like the glass of a greenhouse. That is, it absorbs heat radiation from the earth and from outer space.

Some scientists think that the increasing heat of the atmosphere will cause gradual climatic changes on earth in centuries to come. It is impossible to predict whether such changes will be beneficial or not. If the polar ice caps melt, some coastal areas will be drowned by the rising level of the oceans. But a rise in temperature could also mean that lands too cold to be farmed at present will gradually be opened to farming and habitation.

Q *What is the difference between climate and weather?*

A The climate of any country is the direct result of that country's position on the globe. The North and South Poles, farthest away from the sun, are very cold all year. The equator, where the earth bulges out toward the sun, is always hot. Places in between range from tropical to temperate to tundra. Climatic zones spread in broad, imaginary belts across the globe. There are also local variations in each belt, such as desert and "mediterranean" climates.

The weather in any one place is caused by local features such as mountains and valleys, large bodies of water, warm or cold ocean currents, high-pressure and low-pressure systems, and many other factors that act upon the atmosphere.

Q *What is atmospheric pressure?*

A Atmospheric pressure at any given point is the measurement of the total weight of air above it. Hot air is lighter (that is, less dense) than cold air. In a column of hot air, air molecules expand. They are light in weight and will rise. Thus, a hot column of air exerts light, or low, pressure.

Cold air is heavier (or denser) than hot air. Its molecules are packed tightly together. They exert heavy, or high, pressure on the earth.

Q *What is the biggest, most important feature of Planet Earth—and yet invisible?*

A The atmosphere. It is a vast envelope of gases that surrounds earth for hundreds of miles above sea level. The atmosphere contains the air that all living creatures must breathe; it protects us from deadly cosmic rays; it causes our climate and weather.

The part of the atmosphere where we live is only about 7 miles (11.26 kilometers) high. Beyond that height (the stratosphere) we have to travel in pressurized airplanes or, even farther up, in space capsules, where we are enclosed in a breathable "atmosphere." We must also supply heat for ourselves, for above 7 miles the air temperature may be $-50°$ F ($-45°$ C), with 150 mile-per-hour winds.

At the upper level of the stratosphere, which extends about 15 miles (24 kilometers) above sea level, is a layer of a gas called ozone. This is the vital gas that mixes with and dilutes harmful rays from the sun, yet allows healthful infrared rays to pass through to warm the earth.

Beyond the stratosphere is the exosphere. Here the earth's gravitational pull becomes weaker and rockets need less power to escape into outer space.

Q *What is meant by "prevailing winds"?*

A Prevailing winds are the huge masses of air that continually circle the globe.

   The air in earth's atmosphere is heated by the sun. The warm air rises, but it doesn't escape into space because of earth's protective envelope of atmosphere and also because of earth's gravitational pull.

   The warmest air is formed near the equator. It flows north or south toward the poles. The cold, heavy air at the poles flows north or south toward the equator.

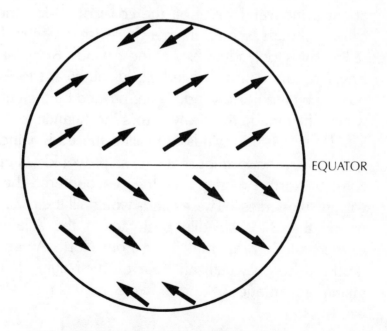

EQUATOR

PREVAILING WINDS AROUND EARTH

If the earth didn't spin, air currents, or winds, would flow strictly north or south. But the earth does spin. It makes a complete turn on its own axis in 24 hours—one earth day. The earth spins in a counterclockwise direction; that is, from left to right, or from west to east. This spinning motion pushes, or deflects, the north-south air masses in a general westerly direction.

Other factors, besides the earth's rotation, affect the air masses and cause localized wind systems to occur. Temperature and air pressure are two major factors involved. To give one example of the effect of high pressure: There is a high-pressure area centered in the Atlantic Ocean, about a thousand miles off the coast of West Africa. It causes air masses to rotate in a clockwise direction. Early sailing ships took advantage of these winds. Ships leaving England, for example, sailed easily southward to Africa, pushed by the north wind. Then, pushed by the easterly winds, they made their way across the Atlantic to America. These easterly winds are called the trade winds.

On the homeward journey, the ships would sail up the coast of North America until they encountered the prevailing "westerlies." These winds would fill their sails and propel the ships steadily back across the Atlantic to Europe. Modern aircraft, flying from North America to northern Europe, gain an hour off their flying time by taking advantage of these same prevailing westerly winds.

Q *What causes thunder-and-lightning storms?*

A A thunder-and-lightning storm is most commonly caused by currents of warm, moist air rising rapidly from the earth into the atmosphere. The updrafts of air meet the cold air of the upper atmosphere. Cold air is much heavier than warm air. You can actually see large, "anvil"-shaped clouds that have risen high into the sky and then become flat-topped. These are typical thunderclouds.

The warm, moist air in the thundercloud starts to fall to earth as rain or hailstones.

Rapid, violent movement of air currents causes electricity to be released. We see this electricity as lightning.

A bolt of lightning heats the air through which it passes, causing the air to expand and then contract again very quickly. We hear the crashing together of air particles as huge sound waves, or thunder.

Q *Is thunder dangerous?*

A No. It is only a sound, though it scares many people and animals. Thunder can do you no harm.

Q *Is lightning dangerous?*

A Yes, very. The electricity released from disturbed air currents can cause trees and houses to burn up. If a bolt of lightning strikes a living creature, it could be killed or seriously burned.

Most houses are protected by metal lightning rods. Metal is a good conductor of electricity. A lightning rod attracts the electrically charged lightning. The electricity striking the rod either dissipates harmlessly into the air or travels down the rod into the ground.

If you ever get caught in a thunder-and-lightning storm, lie down flat on the ground until it's all over. Never, ever take shelter under a tree, no matter how hard it's raining, or how wet you get.

Although wood is not a good conductor of electricity, if the tree is the only tall object around, it will attract lightning. The power of lightning is so great that it can set the tree on fire.

Q *Is it possible for thunder to occur without lightning, or lightning without thunder?*

A No. Thunder is the direct result of the very fast expansion and contraction of air in the path of a bolt of lightning.

However, lightning may occur so far away that you don't see it, even though you hear the thunder. Or you may see faraway lightning but not hear the sound waves of the thunder.

Q *How can you judge the distance of a thunderstorm?*

A By counting the seconds between the flash of lightning and the sound of thunder. Light travels at about 186,000 miles (300,000 kilometers) per second. Sound travels more slowly, at about 1,000 feet (300 meters) per second or about one mile (1.6 kilometers) in a little less than five seconds.

After you see the lightning, start counting seconds at once. If you don't have a stopwatch, say "one thousand and one, one thousand and two," and so on, to count seconds.

If the time between seeing a lightning flash and hearing the thunder is five seconds, the storm is one mile away. If the time is ten seconds, the storm is two miles away. If you see the lightning and hear the thunder at almost the same time, the storm is right over your head.

Q *How is an iceberg formed?*

A An iceberg is formed when a huge piece of ice breaks off the edge of an ice sheet or glacier.

Snowflakes that fell on land thousands of years ago helped form today's icebergs. At the North and South Poles, the "ice caps" of the world, the temperature is so cold that the snow doesn't melt. More and more snow falls until the weight of hundreds of years' worth of snow is packed into solid ice thousands of feet thick. Gradually the ice sheets slip down to the water's edge. Parts of these rivers of ice, or glaciers, break off and become icebergs.

Q *Why are icebergs so dangerous?*

A They are dangerous because although some of them tower hundreds of feet above the sea, nine-tenths of an iceberg is unseen under the water. After the steamship *Titanic* was sunk by an iceberg in 1912, with many lives lost, a constant iceberg watch has been kept to warn oceangoing ships of their presence.

Q *What would happen if the polar ice caps melted?*

A The height of the oceans would rise about 20 feet. Many coastal valleys and cities would gradually disappear beneath the sea. In North America, for example, most parts of New York City, Washington, D.C., Miami, Houston, Los Angeles, and San Francisco would be flooded. But there is no need to panic. Scientists tell us that the ice won't start melting for several thousand years.

Q *What is a rainbow?*

A A rainbow is an arched band of color in the sky. It is formed when light from the sun shines through droplets of water in the air or into a spray of water from a garden hose or a waterfall. Light from the sun seems to be white, but it is really made up of many colors: red, orange, yellow, green, blue, and violet. When light shines into certain objects, such as a prism, or a soap bubble, or a drop of water, the ray of light is refracted (bent) and then reflected by the other side of the water globule. Refraction and reflection cause the light to be broken up into colors.

   The color bands forming a rainbow have red on the outer edge, violet on the inner edge. The colors in between are always in the same order.

Q *What is the major difference between the Arctic (North Pole) and the Antarctic (South Pole)?*

A The Arctic, an area much smaller than the Antarctic, consists of a frozen sea (the Arctic Ocean) ringed by frozen land. The Antarctic, a continent of about 5,500,000 square miles (14,245,000 square kilometers), larger than any country in the world except the U.S.S.R., consists entirely of ice-covered land.

Q *What is the summer solstice?*

A It is when summer officially begins in the Northern Hemisphere, and usually occurs on June 21. The sun's path across the sky on this day is the highest and longest of the year, making its hours of daylight the longest.

Q *Is June 21 the day of the earliest sunrise and latest sunset of the year?*

A No. As sunrise-sunset tables will show, the earliest sunrise of the year takes place about June 4, and the latest sunset about June 27, varying with latitude.

Q *What is the winter solstice?*

A It is the time, around December 22, when the North Pole is tipped farthest from the sun. The winter solstice is the shortest day of the year in the Northern Hemisphere, but it is the longest day and the height of summer in the Southern Hemisphere! Thus, the prevailing conditions are the reverse of those occurring at the summer solstice.

Q *Is there any truth to the saying, "Red sky at night, sailors delight. Red sky in the morning, sailors take warning"?*

A Yes. There may be some truth in it on certain occasions and in certain places. This weather proverb applies principally to the middle latitudes in the Northern Hemisphere, where weather systems tend to move from west to east.

"Red sky at night" indicates that the region to the west, from which tomorrow's weather will come, is free of clouds.

"Red sky in the morning" implies that the region to the east, where the sun rises, is cloud-free, which may mean that the fair weather is receding eastward, making way for clouds approaching from the west.

Q *Is the salty-tasting water of the oceans made up mostly of salt?*

A No. The ocean contains less than one percent of the kind of salt that we use for flavoring. Six kinds of salt make up over ninety percent of the particles dissolved in the ocean: sodium, magnesium, potassium, chlorine, sulphur, and calcium. There are also many other ores, minerals, and metals in ocean water.

Q *Is it true that ocean water contains huge amounts of gold?*

A Yes. But so far, methods for obtaining it are too costly to make a gold rush to the oceans seem likely in the near future.

Q *What is an earthquake?*

A An earthquake is a sudden shuddering or trembling of the earth, caused by shock waves passing through it.

Q *What causes an earthquake?*

A An earthquake occurs when pressures beneath the earth's crust build up to such a point that giant blocks of rocks, pushing against one another, creak and groan (causing shudders on the surface) until eventually one rock plate "rides" on top of the other. A huge rift may be caused on the earth's surface. The impact of these gigantic rearrangements causes shock waves to be felt for miles around.

Q *What was the most destructive earthquake recorded in North America?*

A In terms of property damage the earthquake that destroyed most of San Francisco, California, on April 18, 1906. The San Francisco earthquake, which recorded a magnitude of 7.8 on the Richter scale, affected land all along the San Andreas fault. The Alaska earthquake of 1964 recorded a magnitude of 8.4.

Q *What is the San Andreas fault?*

A It is a huge crack or "fault" (line of weakness) that stretches about 650 miles along the California coast, from San Francisco in the north to the Mojave desert in the south. There are many such faults all over the earth, and it is in these places that severe earthquakes are most likely to occur.

Q *How often do earthquakes occur?*

A About a million earthquakes occur each year! But most are so mild that hardly anybody notices them. Only about a hundred earthquakes per year cause destruction.

Q *How do we know there are so many earthquakes, if people cannot feel them or see any movement?*

A A very sensitive instrument called a seismograph records every tremor of the earth, no matter how small. A seismograph often consists of a large weight that hangs from a framework resting on a fixed base in the earth. When tremors from an earthquake occur, the base of the seismograph moves, while the weight does not. The amount of movement in the base is recorded and traced on a chart, called a seismogram, providing a record of every earth movement.

Q *What is the Richter scale?*

A It is a standard for expressing the magnitude of earthquakes. (The magnitude is the amount of energy of the earthquake at its point of origin.) The Richter scale runs from 0 through 8.9. Thousands of earthquakes are so small that they are recorded as minus numbers on the scale. The weakest earthquakes which might be felt by people are measured around magnitude 2. Buildings may be damaged at magnitude 5. An earthquake of 7 or over is major, and of magnitude 8 or over, a great disaster.

Q *What is the epicenter of an earthquake?*

A The epicenter is the scientific name for the point where earthquakes occur. It is the place where the effects are most severe.

Q *Is there anything good about an earthquake?*

A Yes. If it weren't for earthquakes, which help to form new mountains and valleys, the crust of the planet would eventually be worn down to a flat, swampy plain.

Earthquakes also help scientists to find out something about the earth's interior.

Q *Is the earth made up of one solid ball of rock?*

A No. Nobody has journeyed to the center of the earth, or anywhere near it. But people who study rocks and earthquakes and volcanoes think that the earth is made of several different layers.

The *crust*, or skin, of the earth is the part that we live on. Its thickness varies from 5 miles under the oceans to 20 miles under the land. The temperature increases as we go deeper and deeper into the crust. At 20 miles down, the temperature is 1,600°F (871°C)—hot enough to melt rocks.

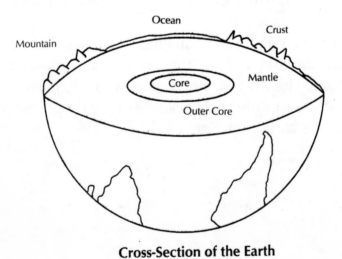

Cross-Section of the Earth

The next layer, the liquid *mantle,* is about 1,800 miles (2,900 kilometers) deep—and very, very hot . . . about 4,000°F (2,205°C).

The *core* of the earth is made up of two layers: The outer core, 1,400 miles (2,252 kilometers) thick, has an average temperature of 6,500°F (3,590°C) and is mostly liquid. The inner core, only 800 miles (1,300 kilometers) in thickness, is thought to be made of solid iron and nickel. The temperature may be as high as 9,000°F (4,980°C).

Q *If the temperature at the center of the earth is the highest in the planet, why don't the rocks in it melt?*

A Because the tremendous weight and pressure of the other layers of the earth squeeze the rock particles so tightly together that they can no more melt than a snowflake could under a ton of ice.

Q *What is a volcano?*

A A volcano is a kind of mountain built up from material that has exploded from deep under the earth's crust. It is always shaped like a cone or a dome. Some volcanoes have taken centuries to be built up. Others grow to remarkable heights in weeks or even days.

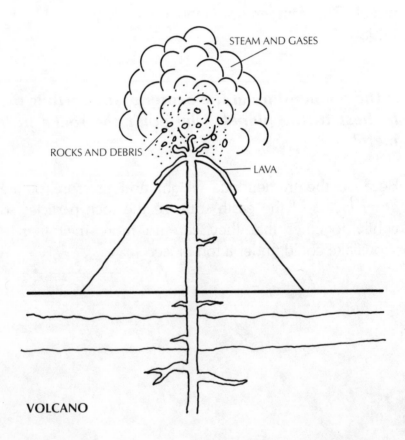

STEAM AND GASES

ROCKS AND DEBRIS

LAVA

VOLCANO

Q *What makes a volcano erupt?*

A The constant movement of materials deep below the earth's thin crust sometimes builds up huge pressures and high temperatures. Gases are formed, looking for a way out. They find their way out through cracks, or "chimneys," that lead from earth's restless underworld to the upper crust. And then there's a big BANG! Flames, clouds of steam, smoke, ashes, cinders, molten *magma* (melted rocks), explode out of the earth.

Q *What is the difference between magma and lava?*

A Magma is the molten rock that lies beneath the earth's crust. When it reaches the earth's surface in a volcanic explosion, it is called lava. Lava is liquid at first. As it cools, it becomes solid.

Q *How does a piece of volcano get into your bathroom?*

A A pumice stone that is used to rub away calluses is formed of solidified volcanic lava. It looks somewhat like gray, solid froth. In powdered form it is used for scouring and polishing many objects, from teeth to furniture.

Q *What was the most recent and best documented volcanic eruption?*

A The eruption of Mount St. Helens in the Cascade Range within Washington State, U.S.A. It occurred at about 7 A.M. (Pacific Daylight Time) on May 18, 1980.

Q *Was the spectacular eruption of Mount St. Helens a big surprise?*

A No. The volcano has been erupting fairly regularly for 4,500 years. Its latest eruption before this one was in 1857. On March 20, 1980 there was an earthquake (magnitude 4 on the Richter scale) underneath the mountain. This was followed by thousands of tremors, over several weeks, which showed that there was great subterranean activity. Scores of scientists and tourists gathered to see the volcano erupt, and some spectacular photographs were taken at close range.

Q *What was the result of the Mount St. Helens eruption?*

A Thousands of acres (44,000 acres) of trees (mostly Douglas firs) were burned to ashes or snapped like matchsticks. There was complete devastation for miles around. Lakes and rivers were choked with mud. Fish and other wildlife were killed. More eruptions continue, and there may be more, perhaps for the next 20 years. Fortunately, the area is not heavily inhabited, but still, some people were killed and houses destroyed. The property damage was about $3 billion. And, of course, the shape and size of the mountain itself was totally changed.

Q *Is it true that continents of the earth are moving?*

A Yes. The world's continents and oceans rest on 12 gigantic plates that are continually in motion. As these huge slabs (each about 60 miles, or 100 kilometers, thick) move, they nudge each other. Sometimes the pressures from within the earth cause the plates to grind against each other and create earthquakes. The San Andreas fault, in California, is an excellent example of the plates in motion. The western side of the fault is part of the Pacific plate. The eastern side is part of the Atlantic plate. The western side is slowly grinding its way northward and westward. The eastern side is lumbering southward. The movement is usually so slow that it's impossible to notice it. But during the San Francisco earthquake of 1906 the fault moved as much as 20 feet (6 meters) in a few minutes.

Q *What is the newest large volcanic island on earth?*

A The island of Surtsey. A volcano rose out of the ocean just south of Iceland in November 1963. At first, Surtsey was tiny. But the volcano kept erupting for months, as scientists, reporters, and tourists from all over the world came to watch its birth. The island now has an area of 1.25 square miles (3.2 square kilometers) and a height of 568 feet (173 meters) above the sea.

Scientists are watching closely as the island is gradually being colonized by forms of plant and animal life, brought there by the sea, dropped by birds, or carried on the wind. From this research and observation they hope to learn more about how our earth developed.

Q *Did the island of Surtsey rise all of a sudden from the sea floor?*

A No, although its appearance above the surface was dramatically sudden. Surtsey erupted from an underwater mountain range called the Mid-Atlantic Ridge. The ridge is the juncture between two gigantic plates of the earth's crust. The volcano that formed Surtsey probably started erupting thousands of years ago, gradually building up a dome that finally reached sea level.

Q *What is snow?*

A Snow is a crystal of ice formed when water vapor meets a microscopic particle of dust in the atmosphere when the temperature is below freezing point. As more ice forms, a beautifully shaped crystal is formed. Many crystals sticking together fall to earth as snowflakes, if the temperature remains below freezing point.

Q *What is known as "The White Death"?*

A An avalanche, which is a cascade of ice, snow, and rocks that falls suddenly down a mountainside. Anything in the path of an avalanche will be destroyed: trees, houses, people, for the force and weight of tons of falling glacial material is as much as that of hurricanes, earthquakes, and floods.

Fortunately, although thousands of avalanches occur, they are usually in remote, uninhabited areas.

Q *What causes an avalanche?*

A In cold areas, such as mountaintops, layer after layer of snow falls, sometimes all year round. If the temperatures are cold enough, the snow doesn't melt. It piles up, getting heavier and heavier. The force of gravity pulls it on a downward path. Under stable conditions, the river of ice, known as a glacier, follows a predictable downward path. Its path has been gouged out by many glaciers before it.

But sometimes sudden changes in temperature, warm rainfalls, earth movements, even man-made effects such as construction blasts, sonic booms from jet aircraft (and, some say, even a yodel or a sneeze), can trigger the delicate balance of the downward drift. The mass of snow and ice may go roaring down a slope like a giant disturbed, and may take unexpected leaps and turns.

Q *Is it true that St. Bernard dogs can find people buried under snow?*

A Yes. St. Bernards and other large dogs that are intelligent and gentle (German shepherds, Samoyeds) can be trained to sniff out people under the snow. However, the dogs do *not* go out on their own, but are led by their trainers. And they do *not* carry little flasks of brandy around their necks!

Q *What causes ocean tides?*

A A tide is a natural occurrence that happens regularly every day because of the moon's gravitational pull on the waters of the earth. When the moon and the sun are on the same side of the earth, there will be exceptionally high tides, called "spring" tides. They have nothing to do with the season of the year.

Q *What is the most destructive sea wave?*

A It is a *tsunami*, often wrongly called a tidal wave. It is caused by some great earth disturbance, such as an earthquake or a volcanic eruption, that has happened underwater. A tsunami can be a single wave or a series of waves, with crests 100 miles (160 kilometers) or more apart. They travel the oceans at speeds of 600 miles (965 kilometers) an hour, and can engulf and destroy coastal buildings over 100 feet (30.5 meters) high. One of the most destructive series of waves in the history of mankind wiped out the lives of over 36,000 people on the coasts of Java and Sumatra in 1883, set off by the mighty eruption of the volcano Krakatoa.

In 1960 the earthquakes of Chile formed a tsunami that traveled 12,000 miles (19,908 kilometers) (nearly half the circumference of the earth) to cause death and destruction in Hawaii, Japan, and other islands in the Pacific.

Q *What was "The Tide of the Century"?*

A It was an exceptionally low tide occurring in March, 1967, in the English Channel. Thousands of people lined the shores and beaches as the tide receded for miles beyond its usual line. Many shipwrecks were uncovered. The low tide was caused because the full moon and the sun were in direct line over the equator, and exerting maximum pull on the ocean waters. This exact alignment happens every 18 years, so the next one should be in 1985.

Q *Where does the highest tidal range occur?*

A In the Bay of Fundy, between New Brunswick and Nova Scotia in northeast North America, about 100 billion tons of water move in 40- to 50-foot tides twice a day. This extraordinarily high tide is caused by the funnel shape of the Bay of Fundy: the ocean water must squeeze through a narrow channel. This phenomenon is called a "bore." Other famous bores occur in the Amazon River (South America), where a wall of water about 15 feet (5 meters) high rushes as far as 200 miles (320 kilometers) upriver before it subsides; in the Ch'ien-tang River, China; in the Solway Firth, Scotland; and in several rivers all over the world where ocean tides must pass through a narrow inlet into a wider basin.

Q *What part of the earth is made up of animal skeletons?*

A The coral reef. There are many such reefs in the world, located around islands and along the shores of continents where the water is warm and clear. Coral reefs are made up of skeletons of tiny creatures called coral polyps. The coral polyp is a soft-bodied animal, but it lives inside a hard skeleton. Its tiny tentacles sweep food into its mouth and stomach. Corals live together in large colonies. Dead corals leave their skeletons behind, which eventually become cemented together and form reefs.

Q *Why are coral reefs valuable?*

A For one thing, coral reefs provide strong barriers between ocean waves and nearby shores. For another, they form calm and quiet waters where many other sea animals, such as sea anemones, sponges, and beautiful tropical fish can live. And, of course, the clear waters of the coral reef are wonderful places for human divers to observe many different life forms.

Q *How are geysers and hot springs formed?*

A They are both formed when water under the earth gets heated by the molten material under the crust. Hot water turns into steam, and steam can produce lots of pressure. The pressure forces the steam to find an escape route— often a place where the earth has cracked during earth movements. Columns of steam and hot water spout up in great jets called geysers.

Thermal or hot springs may take the form of warm lakes. Sometimes underground heat forms mud pots that bubble up in swampy areas. You may see all of these natural wonders in Yellowstone National Park.

PREHISTORIC TIMES

The story of life on earth is very old indeed—much, much older than scientists imagined only a few years ago. Modern methods of probing earth's secrets, of gaining access to remote corners of the globe, of discovering new species and ancient fossils, all seem to push the history of Planet Earth further and further back into the past.

Here are a few questions to excite your curiosity about our ancient world:

When did life first appear on earth? Are there dinosaurs alive today? How can we tell what people ate thousands of years before recorded history? When was the last Ice Age?

Q *How old is the earth?*

A Somewhere between 3.5 and 4.5 billion years old, as far as we know right now. But scientists keep finding new methods of dating ancient rocks, and explorers keep finding "new" ancient rocks; so, apart from the fact that earth is very old indeed, we cannot give it a specific birthdate.

Q *What was the last great catastrophe to leave a permanent change upon the earth?*

A The end of the last Ice Age, about 15-20,000 years ago. The melting of the great ice sheets drowned large areas of land and, by depositing glacial debris, created new land. The passing glaciers permanently changed the look of the land by grinding out great U-shaped valleys, deep coastal inlets called fjords, and great lakes (including the Great Lakes of North America). Plants and animals that had retreated before the glacial flow were now able to re-establish themselves farther and farther north (or south, in the Southern Hemisphere).

Q *When did life first appear on earth?*

A Until recently, scientists thought that life didn't appear until 570 million years ago. But recent discoveries indicate that there were organisms on earth 3.5 *billion* years ago.

It's difficult to think of time in so many millions of years.

Try imagining the earth's existence as a round clock dial, with all 12 hours on its face equaling 4.5 billion years.

We know practically nothing about what happened in the first 2 hours and 52 minutes. The best-known of the early rocks appeared at 2:52 A.M. The first life (bacterial organisms) was well established by 4 P.M., if not slightly before.

Eons dragged by until, at 10:30 P.M., the first vertebrates (animals with backbones) appeared in the seas.

Dinosaurs appeared on the scene at 11:25 P.M., but were replaced by birds and mammals 25 minutes later. Hominids (manlike creatures) arrived at about half a minute before noon. The last *tenth of a second* covers the entire history of human civilization on earth!

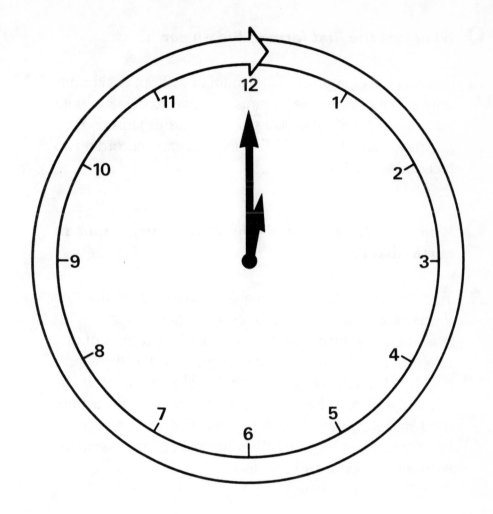

EARTH CLOCK

12 HRS. = 4.5 BILLION YEARS

Q *What was the first form of life on earth?*

A First records of tiny, sea-living animals without backbones (invertebrates) and sea plants such as seaweeds date back at least 550,000,000 years. Some organisms may go back even further (3.5 *billion* years), according to recent discoveries.

Q *What is a trilobite and why is it so important in earth history?*

A Trilobites were the most highly developed of the first forms of animal life on earth, and the most numerous. At least, more of such fossils have been found than of any other animals, perhaps because their bodies were protected by a hard material (chitin) like those of many modern insects. The trilobites ruled the seas for a hundred million years. Scientists think that they may have been the ancestors of modern horseshoe crabs, spiders, scorpions, insects, and shellfish.

Q *How old are the oldest fossils found on earth?*

A About 3.5 *billion* years old. Scientists in 1980 discovered some biological cells of that age in some of the oldest rocks on earth.

Until this discovery, in a remote corner of western Australia, scientists believed that there had been no life on earth until millions of years later. They now think that life may have begun even earlier than 3.5 billion years ago. Keep checking!

Q *How do scientists find out when prehistoric organisms lived?*

A Mostly by a method called "carbon-dating." Carbon-dating is based on the fact that all living organisms absorb a substance called carbon 14 during their lifetimes. When they die, the carbon disintegrates at a steady, known rate. By measuring the amount of carbon 14 remaining in bones, shells, and other remains, scientists can calculate the time when the organism died.

Naturally, there is a limit as to how far back in time a substance can be dated, since at some point there is no more carbon 14 left to measure.

Q *What is a fossil?*

A Fossils are hardened remains of long-dead plants and animals, or their footprints or outlines, that have been preserved in the earth's crust, and sometimes in amber (ancient tree resin), or in other materials, such as coal.

Q *What is amber?*

A It is fossilized resin that oozed from the trunks of trees millions of years ago, in the Cenozoic Era. It may be naturally clear—or it may be opaque, due to the presence of thousands of ancient air bubbles. Sometimes amber may contain the remains of insects and plants of the Cenozoic Era. These remains provide valuable clues to scientists about the kind of organisms that lived millions of years ago.

Q *Why doesn't everything that dies become a fossil?*

A Both plants and animals usually get eaten up by other animals (including microscopic bacteria), or swept away by wind and water. But once in a while an animal or plant may get buried whole, or almost whole, by volcanic ash or mud, or sink into a tar pit, such as the famous La Brea

tar pits in Los Angeles, California. Other organisms may be preserved in desert sands or layers of ice and snow. As the years go by, the layer of earth containing the remains of the organism, or its footprint or leafprint, gradually gets covered with other layers of earth. It may not be uncovered for millions of years, if ever.

Q *Are fossils still being discovered?*

A Yes, almost every day. At first fossils were discovered by chance. It wasn't until the late nineteenth century that scientists began to realize what fossils were and began to study them seriously. Fossil-hunting became a kind of craze, with people putting together all kinds of bones to make "ancient" animals that may or may not have existed in that particular form.

As late as 1975, for example, it was discovered that a favorite dinosaur, Brontosaurus, had the wrong head attached to its body. Now museums all over the world are having to rebuild their Brontosaurus models with heads that are more like those of a close relative, Diplodocus.

Q *Which of today's insects are identical to fossilized species from the Carboniferous period?*

A Only the cockroach, which has survived millions of years without change. The dragonflies of the Carboniferous period resembled those that exist today, but they were not identical. Some had a wingspan of 29 inches (about 75 centimeters). These giants, called meganeurons ("Great Nerve"), were in flight some 100 million years before the pterodactyls and at least 150 million years before birds.

Q *What is the connection between dinosaur bones and storybook giants?*

A Size. When people long ago found enormous bones, they thought the bones belonged to a race of giant people who once ruled the earth. Remember, in ancient times people traveled mostly on foot and, much later, on horseback. They couldn't go very far and there was no way for news to travel fast.

People had never heard of dinosaurs. Most of them had never heard of elephants, let alone huge woolly mammoths. So when they found gigantic bones, they assumed that they had belonged to gigantic people—giants. "Tall tales," or legends about giants, grew up in isolated communities all over the world.

Q *What is the oldest known fossil of a mammal to have lived in North America?*

A As of September, 1981, the oldest known fossil has been dated at 180 million years old. The tiny jawbone of a shrewlike animal was found on a Navajo Indian reservation in northeastern Arizona. The few previous findings of early mammalian fossils in North America dated only to about 135 million years ago. Scientists are excited by the discovery, because it proves that even the earliest, tiniest mammals took many shapes and forms, as they do today.

Q *Why is the hard-shelled egg so important in the history of the world?*

A The development of the hard-shelled, or amniote, egg made it possible for creatures to live on land and be independent of water as a breeding place.

The first life on earth was in the water. Creatures laid eggs that floated in the water. Even when amphibians (such as the ancestors of today's frogs, toads, and salamanders) started to come ashore, they had to return to water to lay their jellylike eggs. Their offspring, like tadpoles, spent their early lives in the water, breathing through gills rather than lungs.

After millions and millions of years the amniote egg developed. Its outer shell is porous: that means it can take in air. It is also hard, compared to jelly: it protects the embryo. Inside the eggshell is the amniotic fluid, in which the embryo floats, and a rich yolk of food on which the embryo feeds.

When the embryo has developed into a baby of whatever species it is supposed to be: a bird, or a snake, turtle, lizard, crocodile—or dinosaur—it hatches out of the egg and is able to take care of itself in a very short time.

Q *What is a dinosaur?*

A The dinosaurs were reptiles that lived about 65 million years ago. We know a little about dinosaurs from studying their fossil remains, and from studying reptiles such as crocodiles, tortoises, lizards, and snakes.

The name dinosaur means "terrible lizard."

Some dinosaurs were indeed "terrible," in that they grew to gigantic size—about 90 feet (27 meters) long and weighing as much as 80 tons (72,800 kilograms). But some were no bigger than today's chickens. And the biggest ones were plant-eaters, posing no real threat to other animals.

Dinosaurs existed in all shapes and sizes, just as today's reptiles vary from a three-quarter-inch Caribbean lizard to 30-foot (10-meter) pythons and giant turtles.

There are two main types of dinosaur: those with hip bones similar in shape to those of lizards, and those with birdlike hips.

Apart from these few facts, we really don't know very much about the dinosaurs. Scientists cannot agree as to which thecodonts ("socket-toothed") were the ancestors of dinosaurs. They are not sure if the dinosaurs were cold-blooded, like today's reptiles, or warm-blooded. There is no way to tell what they really looked like in the flesh, since only bones remain. And we don't know how or why they became extinct.

Keep checking!

Q *What did dinosaurs eat?*

A Some ate plants: they were herbivores. One of the biggest, such as Brachiosaurus, was a herbivore. It must have eaten over 2,000 pounds (about 900 kilograms) of leaves every day to stay alive. Others, such as tyrannosaurs, were carnivores: meat-eaters.

Q *What was the largest meat-eating dinosaur?*

A Its name was *Tyrannosaurus Rex,* which means tyrant lizard king. It lived during the Cretaceous Period, 70 million years ago, in the area that we now know as the Rocky Mountains.

It stood more than 18.5 feet (5.5 meters) tall, weighed about 7.5 tons, and stretched out to more than 45 feet (14 meters) long. Its tail alone was more than 19 feet (6 meters) long.

Tyrannosaurus Rex had about 60 teeth in its mouth, some of them 6 inches (about 15 centimeters) long. They were serrated, just like our modern steak-knives: perfect for cutting meat.

As far as we know at present, Tyrannosaurus Rex was an American dinosaur. But it had many relatives that looked very much like it in other lands, such as that just across the Bering Strait in Asia. The Bering Strait was a land bridge between Asia and North America, thousands of years ago.

TYRANNOSAURUS REX

Q *How do we know which dinosaurs were carnivores and which were herbivores?*

A By the fossil remains of their teeth and their droppings. Carnivores had long, sharp teeth, often with jagged edges, like saws. Herbivores had flat-topped teeth for grinding up plants.

Prehistoric droppings are called coprolites. They can be ground into dust and their contents analyzed.

Q *How big was the biggest dinosaur?*

A The *longest* dinosaur skeleton ever put together is that of Diplodocus, about 90 feet (27 meters) long, including its 50-foot tail and a long, snakelike neck. (That's about the length of two and a quarter public-transport buses lined up!) Diplodocus strolled around Utah 150 million years ago, quietly eating up several tons of plants per day.

The *heaviest* and most *massive* dinosaur was probably Brachiosaurus, who weighed about 80 tons (72,800 kilograms)—that's about the weight of twelve large elephants. Like Diplodocus, it was a peaceful vegetarian.

On swampy ground it left fossilized footprints about the size and depth of today's kitchen sinks.

DIPLODOCUS

BRACHIOSAURUS

Q *Why did dinosaurs become extinct?*

A The extinction of the dinosaurs is one of the great unsolved mysteries of the world. Nobody can answer that question, though many have tried posing various theories. One theory is that the dinosaurs were affected by changes in the earth's climate, which would have been accompanied by a change in the vegetation and possibly a shortage of water.

Another theory says that our planet may have been bombarded by cosmic influences, such as comets or asteroids or cosmic rays that made the earth's atmosphere unbreathable for many plants and animals (but why not all?).

Still another idea is that the continents began shifting at this time, leaving some creatures marooned without sufficient food, while others may have caught diseases from hitherto unfamiliar species of bacteria.

Perhaps mammals helped to kill off the dinosaurs. They were certainly becoming larger, more numerous, and more varied at this time. Many of them probably fed on dinosaur eggs and dinosaur babies.

Perhaps we will never know the truth about the extinction.

Q *How long a time were dinosaurs on earth?*

A About 150 million years, which isn't a bad record, considering that manlike creatures have been around for only 3.5 million years (as far as we know). In fact, dinosaurs were the most successful animals *ever* to inhabit this planet.

Q *Are there any dinosaurs alive today?*

A A few years ago, the answer would have been a definite "No." But in recent years scientists have brought forward studies to show that not all the dinosaurs were cold-blooded. Some of them may have been fast-moving, lively, warm-blooded creatures. These scientists think today's birds are direct descendants of the dinosaurs. Certainly the scaly legs and feet, and the fast head and neck movements, are distinctly reptilian in many species of birds.

Dinosaurs were reptiles—and so they are distantly related to modern lizards, snakes, the tuatara, turtles, and the crocodilians.

Q *Did some dinosaurs look after their eggs?*

A A recent discovery in Montana, in 1981, made scientists think again about the subject of dinosaur parenting. The find was a nesting site of dinosaurs, complete with fossilized eggshells and baby dinosaur bones. The presence of young dinosaurs near the nests has suggested that they were waiting to be fed by the parents, just as with baby birds today.

Another interesting fact is that these are the first eggs of carnivorous (meat-eating) dinosaurs ever to be discovered in the world.

Q *Is it possible that a large dinosaur, such as a sauropod, could still exist?*

A Maybe. In at least one remote corner of the world, in the Ubangi-Congo swamplands of Central Africa, there are persistent tales of a creature as large as an elephant, with a long, snakelike neck and a huge tail. Natives who have been shown pictures of several animals always point to the supposedly extinct sauropod when asked to describe the animal. The region of the sighting has remained relatively undisturbed since the Cretaceous period when sauropods were still living. A group of scientists and explorers set off in the fall of 1981 in search of the beast. Keep checking!

Q *What is the oldest bird fossil ever found?*

A The fossil of Archaeopteryx (ar-kee-OP-tuh-rix) ("ancient wing"). It was found in Bavaria, West Germany, late in the nineteenth century, with the imprints of its feathers clearly perserved in limestone.

Its discovery caused much excitement, for fossils found at the same site showed that Ancient Wing had lived 140 million years ago, at the time of the dinosaurs. In fact, it had many reptilian features, such as a long, bony tail, teeth, and lizardlike claws. Although it had feathers and wings, Ancient Wing probably couldn't fly, for its bones were heavy, and its weak breastbone lacked the "keel" of modern birds. (To the keel are attached the strong chest muscles that move a bird's wings in flight.)

Archaeopteryx may have run along the ground with its wings outstretched to snare insects. It may also have used the wings to beat down small animals before clutching them with its long claws and teeth. It may have climbed up the trunks of trees, digging its claws in, and using its long tail to balance. And then it may have glided from branch to branch, in much the same way as a modern "flying lizard" or "flying squirrel." Its feathers had evolved from reptilian scales and provided good insulation against heat and cold. We do not know if it was warm-blooded, like today's birds.

Q *What was the largest flying animal ever to inhabit the earth?*

A Probably a pterosaur, *Quetzalcoatlus northropi*, which was not a bird but a flying reptile. Part of its fossilized skeleton was discovered in Texas in 1971. It is calculated that its wingspan alone must have been about 24 feet (8 meters), the length of four 6-foot men! (One of today's largest birds, the Andean condor, has a wingspan of about 9 feet [3 meters]).

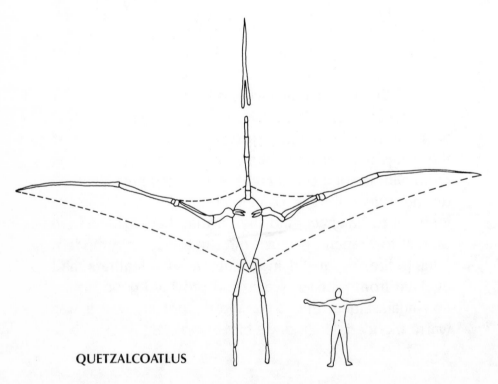

QUETZALCOATLUS

Scientists think that some flying reptiles may have had wingspreads of up to 70 feet (about 23 meters) across, though it is hard to prove because so far most skeletons are too incomplete to make anything but an intelligent guess.

Quetzalcoatlus lived about 70 million years ago.

Q *Who or what was the dodo?*

A The dodo was a real, pigeonlike bird that lived on the island of Mauritius in the Indian Ocean. It was tame and friendly and had no natural enemies until the tropical island was visited by seamen from various countries from about 1598 onward. The flightless, plump bird made good eating and was easy to catch. By 1681—less than a hundred years later—the dodo was extinct. It was made famous by author Lewis Carroll in *Alice in Wonderland*.

There was nothing really extraordinary about the dodo, except that its existence—or non-existence—started people thinking about how quickly creatures could be made to disappear from the earth forever, never to be born again, if they were overhunted.

Q *Are there human fossils?*

A Yes. Scientists from all over the world are competing to find the earliest examples of *Homo sapiens* (the scientific name for modern man) and to trace his immediate ancestors.

Q *Who was the ancestor of modern man?*

A Nobody knows for sure, since scientists are still looking for answers. Besides, they cannot agree among themselves about new fossils and new theories that are brought forward.

Many scientists agree that people as we know them today evolved from earlier forms of apelike creatures. These creatures gradually developed a large brain, erect posture, bipedalism (walking on two feet, rather than four), and, thanks to the opposable thumb common to all simians (apes and monkeys), the use of tools and weapons.

Cro-Magnon man, who appeared about 30,000 years ago, is thought by many to be the immediate ancestor of modern man.

Other people do not believe that man evolved from apelike creatures. They believe that *Homo sapiens* (our species) was created by a supreme being, more or less as related in the Book of Genesis, in the Old Testament of the Bible.

Q *Who or what was the Piltdown Man?*

A He, she, or it was nothing but a hoax—a kind of joke—played on scientists in the year 1912, when everybody was very eager to find the "missing link" between the ancestors of men and apes.

Piltdown is in Sussex, England. Somebody claimed to have found a skull that seemed to fit everybody's idea of what the missing link should look like: part human and part ape. It took years of study to find out that the skull and jaw bones of the "Piltdown Man" came from modern animals, cleverly put together and made to appear aged by some prankster.

The importance of this old hoax is that it made scientists (and newspapers) much more careful about analyzing and publishing "new discoveries" about anything, especially those concerning the human race.

Q *Who, in anthropology, is Lucy?*

A "Lucy" is the name given to the oldest, most-complete, best-preserved skeleton ever found of any erect-walking hominid. It was unearthed in Africa by scientist Donald Johanson in 1974. From the width of the pelvis he assumed that the bones were those of a female. He named the skeleton Lucy and placed her age at 3.5 million years. Her scientific name is *Australopithecus afarensis.*

Lucy is astonishing mostly because she shows us that bipedalism (walking erect on two feet) goes back almost 4 million years. According to Johanson's theories, before Lucy (and her relatives) there were apes. It was Lucy's kind that opened the door to a new species that was to evolve into separate branches: modern apes, creatures that resembled both apes and men, and modern man, *Homo sapiens.*

Like any other major discovery concerning the ancestry of man, there is much controversy about whether Lucy is truly what she seems. It may take many years and many debates—and more skeletons and other evidence—before Lucy is put into her appropriate niche in history.

Q Did prehistoric cavemen draw pictures?

A Yes. Wall paintings of great skill and beauty, and small, carved figurines, called Venuses, have been discovered in many places where our Stone Age ancestors lived, around 35,000 B.C. The most famous are those in Altamira, Spain, and in Dordogne, France. The paintings depict deer, bison, and horses.

Scientists think that the drawings and carvings were related to a belief in magic and the supernatural. Perhaps the artists hoped to inspire good luck during the hunt.

Perhaps, too, the paintings were learning tools for children too young to participate in a hunt.

Q Who were the first Americans?

A As far as we know, the first Americans, or Paleo-Indians, arrived on the North American continent 11,000 years ago. They had traveled from Asia across the Bering Strait, which at that time was a land bridge from Asia to North America. It was about 55 miles long. When the last ice cap melted, the bridge was flooded over.

These migrants were hunters in search of food, hides for clothing, bones for tools. Over thousands of years, these people spread from Alaska to the tip of South America, and from California to Maine.

Q *How can scientists tell what kind of food was eaten by people who lived thousands of years ago?*

A They can tell by "carbon dating" the bones in ancient skeletons. New research shows that certain crops, such as corn, convert the carbon dioxide in the air into plant material based on four-atom molecules. Others, such as manioc, convert carbon dioxide into three-atom molecules.

    Scientists examining old bones can tell fairly accurately, from the amount and type of carbon molecules left in them, how old the bones are, and what their owners used to eat.

Q *Is the modern elephant related to the extinct woolly mammoth of prehistoric times?*

A Yes. The frozen carcass of a young woolly mammoth was found in Siberia (northern U.S.S.R.) in the late 1970s. It was so well preserved (having its entire body, rather than just its skeleton), that scientists were able to make tests on the protein, albumin, and to prove conclusively (in 1980) that this extinct monster was the great, great, very great grandparent of both the African and Indian elephants of today.

    This frozen baby was 40,000 years old.

    The study was made by scientists from the University of California at San Francisco and at Berkeley.

Q *When is a crab not a crab?*

A When it's a horseshoe crab. This amazing creature is an arthropod, belonging to the same general family as spiders and scorpions. Since it has changed very little in the *300 million* years of its existence, it is often called a "living fossil."

Barely visible when born, the almost-transparent creature goes through several molts, when it discards its shell. You may find many of these shells of varying sizes along the beaches of the Atlantic. Adults can grow to several feet in length (including long, spiny tail), weigh about 8 pounds (3½ kilograms), and live more than 15 years.

THE WORLD OF NATURE

Meat-eating plants? Bats that pollinate flowers? White lions? Killer bees, singing whales, trees with knees?

Yes! The world of nature is full of marvels such as these, but we don't have to look for the unusual to become amazed at nature's ways.

A bean sprouting in a glass jar, the song of a sparrow, the brilliant red of a ladybug, a snail's shell—all of these, and much more, are close by for us to wonder and learn about.

Q *Is it true that all animals, including people, develop from an egg?*

A Yes. All female animals produce eggs from which their young will grow, once the egg is fertilized.

Q *Do human females produce eggs?*

A Yes. Human females produce one egg every month from the time they are about eleven years old until they are about fifty years old. If the egg becomes fertilized by a male cell, the egg will grow into a baby, inside the part of the mother's body called the womb, or uterus. The baby takes about nine months before it is ready to be born.

Q *What happens to eggs if they are unfertilized?*

A Unfertilized eggs of female mammals pass out of the body, along with excess blood and the nutrient-rich lining of the uterus. Unfertilized fish eggs simply float in the water until they are eaten by other animals. The unfertilized eggs of domestic chickens are eaten as a highly nutritious and tasty food by people all over the world.

Q *Do all female eggs have to be fertilized by a male?*

A Most of them do, but not all. Among many creatures, particularly insects, there is a process called parthenogenesis (from the Greek words, *parthos,* a maiden, and *genesis,* origin or beginning).

For example, the apple aphid, a tiny but destructive creature, produces swarms of nothing but female aphids all summer long, without the aid of a single male.

Just before winter, however, the aphids produce some wingless males and winged females. The two mate and produce tough-shelled eggs that can withstand the winter cold. When the young aphids hatch out in spring to begin their annual feast of leaves and buds, they are all females. They mature very quickly and start to squirt out thin-shelled eggs that hatch almost at once into hosts of females.

Several other animals, including water fleas, scale insects, and Texas checkered whiptail lizards also reproduce parthenogenetically. Social insects, such as queen bees, wasps, and termites, go on laying eggs all summer after only one mating. They produce mostly females.

The advantage of parthenogenesis is that there is no delay between the generations, and no need to continually seek mates.

Q *Do all females lay eggs the way chickens do?*

A No. There are hundreds of ways in which eggs turn into adult animals. Many insects lay eggs which hatch into larvae; fish lay millions of eggs which float freely in the water before turning into larvae and then fish; reptiles produce eggs which develop either outside their bodies, in nests, or they produce live babies that have matured inside their bodies. Amphibians (frogs, toads, and salamanders) spawn jellylike eggs that become tadpoles. Mammals produce live young after their fertilized eggs have matured inside their bodies.

Q *Why do some creatures produce only one egg, while others produce millions at a time?*

A The number of eggs an animal produces at one time seems to be directly related to the eggs' chances for survival. For example: fish eggs are unprotected. Many of them will get eaten by other animals. A fish lays thousands of eggs at a time. Birds lay fewer eggs, and protect and feed their babies, so they have a good chance of survival. Many mammals produce only one or two babies at a time, and look after them for months or even years. These babies have a very good chance of survival.

Q *Which mammals produce live young that mature in pouches?*

A Marsupials. This is a group of mammals that includes the American opossums, the Australian kangaroos, koalas, wombats, and others.

Q *What is a marsupial?*

A It is a kind of mammal that is born before it is completely developed. Instead of completing its growth inside the mother's womb, as other mammals do, the marsupial baby crawls into a pouch on the mother's body. The journey to the pouch is a difficult one, especially when the baby of a kangaroo, for example, is no bigger than an inch (2.5 centimeters). It must find its way through the mother's belly-fur and into the pouch. Once safely there, it attaches itself to a nipple, and there it stays until its limbs and other organs are fully developed and its own body is protected by fur. Even after it is grown (about eight months, for a great gray kangaroo) a baby kangaroo, or "joey," may hop back into its mother's pouch for shelter for about another six months.

Q *Is it true that some plants are meat-eaters?*

A Yes. There are over one hundred species of the Drosera family that eat insects. The best known are the sundews and the Venus's-flytrap. In natural conditions they grow in water-logged places such as swamps. Water prevents their roots from getting the oxygen and nitrogen that they need to survive. Insect-eating plants find a way around this difficulty by attracting and devouring insects, which provide them with protein (which contains oxygen and nitrogen).

Q *How do plants "eat" insects?*

A First, the insect must be attracted to the plant. The leaves of the plant are hairy and sticky and sparkle in the sunlight (hence one name, sundew). Once an insect lands on a leaf, the hairs or tentacles fold over the creature, holding it fast. Digestive juices from the plant rapidly change the victim into a kind of soup that is easily absorbed by the plant. Then the hairs straighten up, and the indigestible parts of the insect (wings and shell) are blown away by the wind.

Experiments have shown that meat-eating plants will feed upon raw hamburger or particles of hard-boiled eggs (protein) but will not be fooled by pebbles or other inorganic, non-protein fakes.

Q *What is nectar?*

A It is a kind of sugar solution produced by plants. Its chemical composition and quantity varies a great deal from plant to plant, which is why there are so many different kinds of honey. (Bees make honey from nectar and pollen.)

In the legends of ancient Greece nectar was the name given to a drink that was so delicious that it could overcome death and make the drinker immortal. When early botanists (scientists who study flowers) found that bees obtained a liquid from plants and turned it into sweet and delicious honey, they called the plant-produced liquid "nectar."

Q *Why do plants produce nectar?*

A Nectar is one of the many ways in which a plant attracts insects (such as honeybees), birds (such as hummingbirds), and bats. Showy petals, hundreds of different colors, lines and speckles that look like "landing strips," and odors that range from what people consider superb (roses, for example), to unpleasant (such as skunk cabbage), are other methods of attracting creatures that will spread pollen from one flower, or one plant, to another to help make new plants.

Nectar oozes from glands in the plant called nectaries. The nectaries can be situated almost anywhere in the plant. They are carefully hidden so that the pollinating insect must reach them along a well-defined path and touch the parts where pollination can take place.

Q *If some plants can reproduce without pollination, why do they have pollen-making flowers?*

A Because it is an advantage to have more than one way to reproduce. If, for example, there are long periods where conditions are unfavorable to a particular kind of plant (a shortage of insects, perhaps, or too much or too little rain), the plants that can reproduce without pollination can still survive, while plants with no "back-up" system may die out.

The advantage of pollination, on the other hand, is that a pollinated plant has a mixture of genes from male and female parts of plants, and stands a better chance of at least some genes in the mixture adapting to changes in the environment.

Q *What is pollination?*

A Pollination is the fertilizing of a female plant egg (an ovule) by a male cell produced by pollen. Pollen looks like yellow dust. It is formed in the male part of the flower called the anther.

SECTION OF BASIC FLOWER

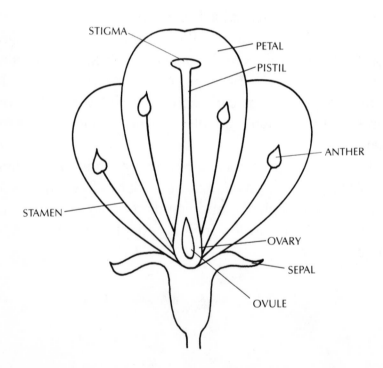

STIGMA

PETAL

PISTIL

ANTHER

STAMEN

OVARY

SEPAL

OVULE

When a visiting insect (such as a bee or a butterfly), or a bat or a bird (such as a hummingbird), brushes against the anther in its search for pollen and nectar, some of the pollen clings to the animal's body. The pollen gets carried to the female part of another flower or to the female part of the same flower. Grains of pollen get caught on the sticky tip of the female stigma. The pollen (a microscopic but remarkable cell) then develops a tube that makes its way down to the ovary ("seed box") of the flower. In the tube, a male sperm is formed. It unites with an egg in the seed box, and a fertile seed begins to grow.

Q *How do plants reproduce?*

A Different plants reproduce in different ways. The plants that we know best, such as most of the wild flowers, weeds, and garden flowers, and trees and shrubs, reproduce by pollination.

Q *Do all plants need the help of animals to reproduce?*

A No. Some plants, such as grasses and many trees, rely on the wind to carry pollen. A few, such as certain pond-weeds, have their pollen carried off in the water.

Many plants do not need to be pollinated to reproduce. Strawberries, for example, produce "runners" that travel from the parent plant along the ground, sprout roots, and pop up as plants. Many "bulb" plants, such as the crocus, or tubers, such as potatoes, divide underground or send up shoots from the parent plant.

Q *How do bats pollinate flowers?*

A With the help of a very long tongue! In one species the tongue is 76 mm (2.864 inches) long, while the body is only 80 mm (3.120 inches) long. The tongue ends in a kind of hairy brush. The bat may be able to collect nectar by hovering, like a hummingbird. But usually it grasps the flower with its clawlike "thumbs," then sticks its head inside the flower. The claws leave marks, so that we know which flowers have been pollinated by bats, who feed at night and are hard to see.

A bat is much larger than any insect. The flowers on which it feeds are correspondingly larger than most: they are usually night-blooming tropical flowers, such as the cup-and-saucer plant *(Cobaea scandens),* the flowers of the banana, sausage, and baobab trees, and others. These flowers produce a great amount of nectar and pollen for the hungry bat, and are sturdy enough to bear its weight.

Q *Once the plant has produced seeds, how does it get the seeds distributed?*

A "By land or sea or air"—or with the help of animals.

Some seeds are spread by the wind (dandelion seeds, for example); some by water (coconut seeds, for example, that are heavily protected but still light enough to float across thousands of miles of ocean). Some seed pods, such as those of the jewelweeds and the witch-hazel tree, "explode," shooting out seeds.

Some plants produce tasty fruits such as cherries or berries. Animals eat these fruits and discard the seeds far away from the parent plant. Other plants produce seeds that hook onto an animal's fur (the teasel or cockleburr, for example) and get carried for miles.

Squirrels are well known for hiding stores of nuts from trees. If these nuts don't get eaten by the squirrels or other animals, they may grow up to be trees.

Q *What makes a flower smell?*

A Tiny particles, commonly called scent strips, are programmed by the flower to give off a scent that will be
attractive to a pollinating insect. The scent strips are
usually present in the petals of a flower, but other parts
may also produce odors of varying intensities. The general idea of scent is to make an insect's path to the
flower's pollen a pleasant and easy one.

It is very difficult to describe a scent or to say why you
like it or why you don't. Fortunately for plants, most
insects and other pollinating creatures seem to have a
"memory" for scents. They know what they like and are
attracted to it.

Q *How do people use a flower's scent to make perfume?*

A One of the commonest ways is to place the petals of a
scented flower onto plates of glass that are thinly coated
with grease. The grease absorbs the scent, which is then
extracted from the grease in the form of a highly concentrated oil. The oil will be mixed with other chemicals to
make perfume.

Q *Does a millipede (whose name means "thousand feet") really have a thousand feet?*

A No. No species of millipedes have more than two hundred legs. Some have only eight legs. However, some of the millipede's ancestors (among the first animals to leave the water and live on land) were as long as cows! They probably had at least a thousand feet, one pair to each segment of the body.

Q *Why and how do spiders spin webs?*

A Spiders (which are not insects) spin webs to trap the insects on which they feed. Spiders have a remarkable little mechanism, called a spinneret, at the tip of the abdomen. The spinneret produces a silky, hair-thin thread which dries on contact with the air and becomes relatively strong and elastic. It is also sticky enough to catch an insect's feet or wings.

The insect's struggles to escape create vibrations throughout the web and alert the hidden spider, signaling that dinner is ready.

With their spinnerets, spiders also make silken linings for their burrows or cocoons for their eggs. Baby spiders are often carried aloft on wind-borne gossamer and land many miles away from their birthplace. Spider silk is used in certain optical instruments.

Q *What is an albino?*

A An albino is an animal totally lacking in normal pigmentation (coloring). Its skin, hair, or feathers, and the iris of the eye, are white. The eyes appear pinkish because of the color of the blood vessels. Albinism (the condition of an albino) is an inherited characteristic and very rare in nature. However, it has been bred into certain animals, such as mice and pigeons.

Q *Is there such a thing as a white lion that is not an albino?*

A Yes. Until 1975 the answer would have been, "No." The natural tawny color of the lion blends with the dry grasslands where it lives. But in 1975 naturalist Chris McBride witnessed the birth of two white cubs into a pride of normal, tawny-colored lions. The eyes of the lion cubs were not albino-pink, but lion-yellow and perfectly normal. Then, a year later, another white cub was born to another mother. Nobody knows what is causing this new color strain.

The place was Timbavati, a wildlife reserve in South Africa.

There have been legends of white lions for centuries. This is the first time that white lions have been photographed and studied in detail. Keep checking!

Q *What is a "killer bee"?*

A It is a species of African bee with a sting toxic enough to kill a person. In 1956 some of these bees were imported into Brazil to breed with regular honeybees. It was thought that their offspring would produce more honey. But some of the African bees escaped. They started to multiply and move across the country. By the early 1970s 300 people had been stung to death by these "killer bees." Scare-stories about these bees began spreading up to North America. They told of thousands of people being killed.

But the African bees started to mate with native honeybees and now are producing offspring that are not dangerous (except to people with an allergy to bees).

Q *What animal has teeth all over its body?*

A The shark. Instead of having scales, like other fish, the shark has a very tough covering that is studded with *denticles*. Denticles are small, hard bumps that are true teeth, each with a central pulp canal containing nerve and blood vessels. Rubbed the wrong way, this rough hide can scrape ordinary skin raw. In fact, sharkskin was long used by furniture makers to "sandpaper" the toughest woods.

Q *How does a snake, without arms or legs, get its victim into its mouth?*

A First it kills or stuns its victim. A constrictor will suffocate its prey. A venomous snake poisons it. When the victim is dead, the snake seeks its head (the pointed end), and grips it firmly in its jaws. The teeth of the snake keep a firm hold on the meal, while the upper and lower jaws of the snake move alternately to push the victim down the snake's throat to be swallowed.

SNAKE JAWS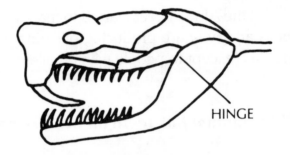

HINGE

Q *How does a snake shed its skin?*

A It crawls out of it, peeling it off as you would peel off a tight turtleneck sweater, so that it comes off inside out.

A few days before shedding, the snake's body produces a kind of oil which covers its body between its two

outer layers of skin. At this time the snake (even its eyes) looks dull in color and its eyes look slightly bluish. It may be irritable, for it is now partially blind and helpless.

When it is ready to shed, the snake may open its mouth in a wide gape, like a yawn, to loosen the skin around the lips. Then it rubs its snout against a rough tree bark or rock to loosen the skin. More rubbing helps the skin to peel back over the head and over the entire body.

The newly shed skin is inside out. It is moist and soft and can be stretched out to its full length. If a pet snake has shed its skin, you will clearly see the pattern of the scales, but there will be little color, just as there is little color on a piece of peeled, sunburned skin. The "new" skin on the snake's body, however, will be shiny and well marked.

Q *Do snakes have teeth?*

A Yes. They have six rows of them—four rows along the jaws and two across the roof of the mouth. The teeth are not deeply socketed like those of mammals, so they come out easily. They are replaced almost immediately by new teeth, throughout the life of the snake. The teeth slant backward, so they are useless for cutting or grinding, but they do allow the snake to get a firm grip on its victim while it is swallowing it.

Q *A snake has no cutting or grinding teeth with which to cut its meal into small pieces. How can it swallow an animal bigger than itself?*

A The skin around a snake's mouth and throat is as stretchable as elastic, and its jaws are not attached to each other. That means that each jaw can move separately and alternately. The jaws are attached to the skull in such a way that they open to a very wide gape. Even the bones of the snout are not firmly fixed together.

Q *Do whales "sing"?*

A Yes, but not in the way people do. They do not have vocal chords. Their sound-making mechanisms are a complex system of air sacs and tubes on either side of the blowhole. Some of their sounds are so high-pitched that they cannot be heard by human ears except with the aid of hydrophones (underwater microphones) and sloweddown tape recorders. Recordings have been made of their "songs," and thousands of people have listened to these chains of sound that are remarkably varied. The sounds range from squeaks and whistles to creaks, groans, quacks, and eerie moaning.

Q *How did the right whale get its name?*

A Because, according to whalers, it was the "right" whale to kill. It was slow enough to catch, fat enough to float after the kill, and carried in its jaws a treasure in whalebone.

For some reason, this creature seemed unafraid of man, and came close to shore, making it easy to hunt, even from small boats launched from shore. So many "right" whales were killed that the animal became almost extinct. The 3-4,000 remaining *Baleana glacialis* are now fully protected. This whale is about 50-60 feet (15-16 meters) long and weighs about 40 tons (36,000 kilograms).

Q *How does an anteater avoid getting bitten and stung by ants and termites?*

A It moves fast! Instead of settling down to a leisurely meal at one termite mound, it snacks at several different ones. First it digs a hole in the mound with its strong, curved claws. Then it darts its 22-inch tongue in and out of the hole at a rate of about 160 times a minute. By the time the fierce soldier ants or termites have rallied forces, the anteater is on its way to the next fast-food stop.

Q *Why do whales sing?*

A Like other animals, whales use sound to communicate with each other. Their sounds carry for several miles underwater.

Dolphins (small whales) in captivity have been closely observed by scientists, who have listened to dolphins "talking" to each other, even when in separate tanks. They have heard a more experienced dolphin giving "instructions" to a newcomer who is being taught to do tricks by its trainer.

Efforts have been made to teach dolphins to speak "human" language, so far without success, though it is obvious that the dolphins are highly intelligent, friendly, and cooperative. They seem to understand what is required of them and some have mimicked human sound patterns.

Keep checking!

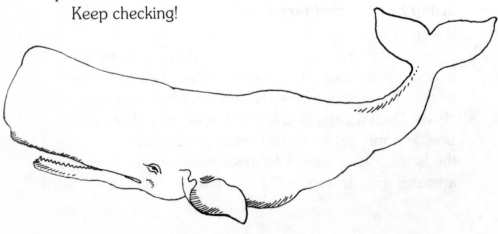

Q *Are jellyfish really fish?*

A No. They have no backbone. They are invertebrates of the large coelenterate family that also includes sea anemones, corals, and the Portuguese man-of-war.

Q *Are jellyfish dangerous?*

A Some of them are. The sea wasp or box jellyfish *(Chironex fleckeri)* is dangerous to all animals, including people. It inhabits the surfs of northern Australia and Indonesia. Its jelly sac is about the size of a human head. Its tentacles are as long as 20 feet. Its toxin is so powerful that it paralyzes the heart muscles and causes death within five or ten minutes.

All jellyfish release toxin (poison) to stun or kill their prey, usually small crabs and fish. The moon jellyfish *(Aurelia aurita)*, common in waters all over the world, is practically harmless. At most it causes a skin rash that may last a few hours.

Q *Do fish make noises?*

A Yes. Through modern techniques of listening and recording, as well as from stories told by undersea divers, we know that fish make all kinds of beeps, squeaks, and grunts. In fact, there is a small coral-reef fish that is called the grunt. Grunts make the grunting sound by grinding together special teeth located in the throat.

Q *Can living creatures exist where the water pressure is about 1,000 tons per square foot?*

A Yes. In 1951 Dr. Anton Brunn's Danish ship *Galathea II* brought up a sample of the sea bottom in the Mindanao Trench—an abyss almost 35,000 feet deep, where it is completely dark, with near-freezing temperatures. The sample came from 33,460 feet and to everyone's amazement it contained bacteria, sea anemones, mollusks, and a minute crustacean. One of the mollusks, a cup-shaped ancestor of the clam, was thought to have been extinct for 300 million years. It was given the name *Neopilina galathea.* Somebody remarked that the discovery of this lowly creature was as astonishing as if a Brontosaurus had come lumbering out of an unexplored jungle today! Keep checking!

Q *Which fish "shoots" and drowns its victims?*

A The archerfish. Waiting near the surface, it "fires" beads of water at an insect by closing its gill covers and forcing water out of a tiny tube in its mouth. The insect is stunned by the water "bullets," falls into the water, drowns, and is quickly swallowed by the archerfish.

Q *What fish looks a little like a horse and acts like a kangaroo?*

A The seahorse. This little creature, with a head shaped like a horse's—it is only about 6 inches (15 centimeters) long—carries itself upright, often curling its tail around a seaweed when it wants to stay still. Most extraordinary of all, the mother seahorse lays her eggs in the father's kangaroo-like pouch, then swims off and leaves him to take care of the babies. The babies hatch out after about four weeks. The father seahorse stays around until the babies have grown to their full length and no longer need his pouch as a harbor.

Q *What fish goes fishing complete with line and bait?*

A The anglerfish. This deep-sea fish looks like the rocks among which it hides. A long, wiggly line extends out of its mouth, at the end of which is something that looks exactly like a tiny fish. It even has spots that resemble eyes. Other fish are easily fooled by this remarkable fake, try to swallow it, and are quickly sucked into the angler's mouth.

Q *What are "mermaids' purses" sometimes found along the seashore?*

A They are the tough, leathery capsules in which skate eggs are hatched.

Skates, like rays, belong to the same family as the sharks. Skates and rays are bottom dwellers. Their bodies are flattened and their side fins widened into what look like wings. Although many are large, all but the stingray, which has a poison barb at the end of its tail, are harmless to people.

Q *Is the devil ray a dangerous creature?*

A No. In spite of its name and its huge size (it can have a "wingspread" of about 22 feet—7 meters—and weigh about 2,000 pounds—900 kilograms), this is one of the gentle giants of the sea. Many divers have swum alongside or even on top of these manta rays, which are harmless to people. They feed on small fish and plankton.

Q *What animal has eyes at the end of its arms, and feet under its arms?*

A The starfish, which is more correctly called a sea star, because it isn't a fish. (It is an echinoderm, part of a large family that includes sea urchins, brittle stars, and sea cucumbers, to name only a few.)

A sea star's eyespots don't work like ours: they can sense only light and darkness, which is all that is necessary on the ocean floor on which the creature lives.

On the underside of the sea star's arms are rows and rows of small, tube-shaped feet. These feet act like suction cups to help the starfish move around, cling onto rocks, and capture prey.

Q *How does a starfish (or sea star) get a clam out of its shell?*

A The starfish wraps itself around its prey, attaches its suckerlike feet, and pulls. It may have to hang on for hours before an opening appears between the hinged edges of the clam. When the opening appears, the starfish turns its stomach inside out, slips it into the clamshell, and digests the clam.

Q *What is the world's tallest animal?*

A The giraffe, a native of Africa. Its height can be over 16 feet (5 meters). An average adult weighs between 1,200 and 4,000 pounds (550 to 1,800 kilograms). It eats about 30 pounds (14 kilograms) of leaves a day, and lives to be about 28 years old.

Q *Does a giraffe have more neck bones than any other animal?*

A No. It has seven neckbones (or cervical vertebrae), the same as deer, cows, and many other animals. Its neck bones are, however, longer than those of most animals.

Q *Why does the giraffe have such a long neck?*

A Nobody knows all the answers to questions about why any animals developed as they did. But one reason must certainly be that the evolution, whatever it is, helps an animal to survive in its particular environment.

In the case of the giraffe, it seems likely that its long neck allows it to reach the topmost leaves of trees, while smaller animals eat the lower leaves. Its long neck also lets it have a good view of the landscape so that it can see approaching predators and take flight on its long legs.

Q *What animal can walk over a razor's edge without getting cut?*

A A snail. As it travels, a snail puts down a path of slimy mucus that protects its foot and aids its movement. The foot is the part of the snail that rests on the ground. You may sometimes see a snail's trail (dried mucus) on rocks or on sidewalks—or even on razors.

Q *Do bats suck blood from other animals?*

A Vampire bats suck blood, usually from cattle, hardly ever from people. These bats are dangerous because they may carry the rabies infection from one animal to another.

Vampire bats live in tropical and subtropical America.

Q *Are all bats dangerous?*

A No. Most bats feed on fruit and insects. They help to keep the insect population down, and are friends, rather than foes.

Q *How do bats find their way in the dark?*

A By *echolocation*. Bats send out constant sounds that are too high-pitched for most animals to hear. The sound waves bounce back from any object in the bat's path and give it an accurate "picture" of the object. One of today's marvels of technology, radar, is based on exactly the same principle as the bat's echolocation.

Q *Is a bat a bird?*

A No. It is a mammal. It suckles its young just as human mothers and other mammals do.

Q *Are there any other flying mammals, besides the bat?*

A No. Some animals, such as "flying" squirrels, seem to fly. As they leap from tree to tree, flaps of skin along their sides help to keep them airborne. But no mammals except bats have wings.

Q *Do insects have blood?*

A Yes. But an insect's blood is almost colorless instead of bright red like that of a mammal. That's because insect blood doesn't have the red-colored cells (hemoglobin) that mammals have.

  If you squash a mosquito on your arm and red blood squirts out of the insect, it's because the mosquito has just had a meal from a red-blooded creature—perhaps you.

Q *What does all animal blood have in common?*

A It carries particles of minerals, food, and salts from one part of the body to the other, and helps get rid of wastes.

Q *Is it true that an owl can turn its head in a complete circle?*

A No. However, it can rotate its head to about a 270° angle, and then swivel it around to the other side so quickly that it seems to have moved in a complete circle, which is 360°.

Q *Do birds have ears?*

A Yes, one on each side of the head. But they don't have outer "funnels" to help collect sounds the way the ears of mammals do. Instead, they have small holes, which are easiest to see on naked-headed birds like vultures.

   The "ears" of birds such as the long-eared owl are merely extra-long tufts of feathers situated on each side of the head, where we would expect to find ears.

Q *Which birds have the fastest wing-beats?*

A Hummingbirds. They are the tiniest feathered creatures in the world. They are well known for the beauty of their plumage (feathers), for their size (the smallest is about 2¼ inches long), and for their remarkable flight.

   A hummingbird's wings move so fast that it is impossible to see anything other than a blur when they are in motion. Slow-motion pictures show that their wing-beats may be as fast as 50 to 75 times *per second.*

Q *Besides its speed, what else is remarkable about the hummingbird?*

A The hummingbird is one of the few creatures that can move forward, backward, or even seem to stay motionless in flight. This particular adaptation is useful because the hummingbird feeds on plant nectar like a bee. It has to hover near a flower, pushing its long beak into the flower to suck up the moisture. Its fast-whirring wings help it to stay in place long enough to get nectar from any flower.

Q *Why do many birds fly in V-formation?*

A The V-formation seems to be a survival technique. As a bird flaps its wings, it disturbs the air and leaves whirling eddies of air behind it. Birds traveling together can take advantage of the upward swirls of the air created by stronger birds ahead of them.

Q *Is a bird an animal?*

A Yes. According to scientists, everything in the world is placed in one of three categories: animal, vegetable, or mineral. A bird is a living, breathing creature that can move around. It is definitely an animal.

Q *What makes a bird different from most other animals?*

A Its feathers. This answer may surprise you. Perhaps you thought the answer would be: a bird can fly. It's true, most birds can fly. But so can most insects, and there are more insects than birds (or any other animal) on this planet. And there's also a mammal (the bat) that can fly! Only birds have feathers.

Q *How does a bird fly?*

A With its wings, of course, but there's much more to flying than having wings. A bird's bones are very light—full of air pockets. In fact, its entire body is full of air sacs. Its pectoral (chest) muscles are very large and strong: they are the muscles that move the wings. (Try flapping your arms up and down a few times—it will help you to imagine the muscular power needed for the bird's "arms" (wings) to carry it upward.) The bird's feathers, of different shapes, sizes, and textures, also play an important part in keeping a bird aloft. It uses its tail as a rudder.

Q *What natural forces help a bird to fly?*

A Wind or hot air (thermal) currents. Birds, especially large-winged species such as sea gulls and hawks, make use of air currents and float on them, their wings motionless. (Human hang-gliders and free-fall artists use those same air currents to travel in the sky, supported by air.) Riding on air currents is called soaring.

Q *What was the Black Death of the Middle Ages?*

A It was a terrible disease, the bubonic plague, which was spread throughout all of Asia and Europe by rats. The plague (which still exists) is spread by fleas that live on rats. Rats travel on ships and on land vehicles, and are probably airborne, as well, hidden in airplanes. Wherever people travel, rats travel, too. The bubonic plague killed off one-third of the world's population in the fourteenth century—the greatest catastrophe to the human race in history.

Q *What animal is considered Public Enemy Number One?*

A The rat. As long as people have been around, the rat, cunning and adaptable, has stayed close. It destroys food, infects and kills animals, gnaws through wood and wire, and worst of all, spreads terrible diseases.

Q *Did the unicorn ever exist?*

A As far as we know, the *unicorn* (a word which means one-horn) didn't exist as a land animal. However, there is a "one-horned" whale, the narwhal. This small whale lives in Arctic seas, and though it has been hunted for centuries, scientists know very little about it. Its single "horn" is really a tooth that grows in the upper jaw of an adult male. It can protrude 7 or 8 ft forward from the upper lip, its surface grooved in a left-handed spiral. It may use the tusk to fight off other males, or to stir up prey from the ocean bed, or to poke holes in the ice.

Q *Why do some kinds of cypress trees have "knees"?*

A To help them breathe. The bald cypress, or swamp cypress, *(Taxodium distichum)* grows from Chesapeake Bay southward. It is often found in swampy areas, where its base and roots are covered with water. Surrounding such a tree you will see what look like small stumps, or "knees," all around the base of the tree. These are upward extensions of the roots. They help get enough air to the roots to prevent the tree from drowning.

Q *What is the world's tallest tree?*

A The redwood, a kind of sequoia *(Sequoia sempervirens)*. These trees grow in northern California and in Oregon. The tallest are over 250 feet high, with the first branches starting at about 100 feet up, so that all you see at first is a tall, bare column. The tallest trees started growing thousands of years ago, perhaps while Julius Caesar was ruler of the Roman Empire, and while the pyramids of Egypt were being built. These are healthy trees, and botanists see no reason why they shouldn't go on living for ages. Sequoias grow naturally only in California and Oregon, though some have been successfully transplanted to other states.

Q *What is the world's oldest tree?*

A The bristlecone pine *(Pinus longaeva)*. Like the redwood, it is found in California. Scientists estimate that some of these trees are at least 4,600 years old and may live to be about 6,000 years old.

Q *What is the world's largest flower?*

A The rafflesia, found on the island of Sumatra. It was named after Sir T. Stamford Raffles (1781–1826), a British governor of Sumatra. But he can't have been very pleased at having his name used, for the flower, when it blooms, is about three feet (one meter) across and smells like rotten meat! The rafflesia plant grows underground, taking food from the roots of other plants. When the time is right, flower buds that look like huge brown cabbages sprout up and smell!

Q *Which mammal has the greatest weight difference between mother and newborn baby?*

A The giant panda. A baby giant panda born in captivity in Mexico was crushed to death by its 300-pound mother shortly after birth. The baby weighed about 12 ounces.

Q *Why is there such a weight difference between adults and babies in some mammal families?*

A Nobody knows. Other animals with a huge weight difference are the marsupial babies, such as kangaroos, which are born before they are fully developed, and bears (not related to pandas). Since mother bears give birth when they are in a sleepy state that is almost like hibernation, many bear cubs are accidentally crushed to death by their large mothers.

Q *Do leaves change color in the autumn?*

A The leaves don't really change color. As the days grow shorter, or during times of drought or disease, the tree reduces its production of the green pigment chlorophyll. The lack of green coloration, which is usually so dominant, allows other colors, such as yellow and red, to show through clearly.

Some of these colors have existed in the leaf all along. Other colors develop as a reaction to increased amounts of sugar trapped in the leaf.

Q *Why do leaves fall from trees in autumn?*

A Because the tree is "drawing in" its life-giving sap to help preserve itself during the winter months and to aid in the formation of new buds. Gradually, the place where the stem of the leaf joins the branch becomes sealed off. A gust of wind is enough to separate the leaf and make it fall. A tiny scar is left on the branch where the leaf used to be.

For some reason, not clearly understood, leaves such as those of the European beech may stay on the tree all through the winter, withered and brown.

Q *What is hibernation?*

A Hibernation is a way of surviving the cold of winter for some animals such as ground squirrels, woodchucks, and bats.

These animals eat all summer and store fat on their bodies. When winter comes, they find a shelter and settle down to "sleep." Hibernation sleep is not the same as our nightly rest or occasional naps. The body temperature of the hibernating animal drops, the heartbeat slows, the animal seems scarcely to breathe. It is using almost no energy during this deep sleep.

Other animals that go into the same state during hot, dry periods are said to be "estivating." (A lungfish, for example, forms a mucus-lined cocoon in the mud and "sleeps" until the rains come.)

People often refer to bears as hibernating, but in fact bears are up and around quite often during the winter in fine weather. When it is very cold, however, they do take shelter and fall into a deep sleep similar to that of true hibernation.

ANIMAL BEHAVIOR

We all know that most mammals and birds are excellent mothers. In some species, such as wolves and penguins, the fathers share parental duties. In others, such as bears and cowbirds, the fathers disappear very rapidly.

But did you know that in the world of fishes, frogs, and toads, there are some amazing fathers that take full responsibility for the young?

People have come a long way from studying animal behavior in order to hunt them (or avoid them) to present-day concern about trying to prevent their extinction in their diminishing environment on our planet.

Can we teach apes to talk to us? Do animals have built-in magnetic compasses in their brains? Do whale babies have aunts? Find out the answers to these and many more questions about our fellow earthlings.

Q *Why is it so important to study the behavior of animals?*

A There are several reasons. The oldest reason for man to observe animals is survival. Once, all men had to hunt animals for food—and they had to flee from certain animals to avoid being killed. Hunters must know a lot about the habits of both prey and predators. Some tribes still depend upon hunting for their existence.

Later in history, some peoples learned to tame certain animals and breed them in captivity: cattle, horses, and dogs were among these domesticated animals.

Today, scientists study animals such as apes, monkeys, and rats to try to understand how their organs work, why they behave as they do, how their bodies react to certain chemicals. Most new drugs are first tested on animals to find out if they will be safe for humans. Many new operations, such as heart transplants, were first tried out on animals.

Zoologists study animals to help the animals to survive. They tag the animals to find out where the animals travel, how much space they need, what and when they eat, when they breed, and how they rear their young. By finding out the needs and behavior of animals, scientists can help them to survive in a world that has less and less room for wild animals.

Q *What is a radio collar?*

A A radio collar looks just like a dog collar, except that it contains a tiny transistor radio that emits a constant signal.

Q *Why do scientists attach radio collars to wild animals?*

A It's done to help them keep track of the animals. We know very little about the habits of animals in the wild. By listening to the signals transmitted by radio collars, scientists can learn the distance an animal may travel in a day, when it hunts, when it sleeps, and how often it eats. It's important to have this information if we are to help wild animals survive.

A recent study on Bengal tigers showed that each tiger needs a hunting ground about the size of five thousand football fields! A good tiger reserve has to be very, very large for even ten tigers. And, of course, it has to be protected from hunters.

Q *What is survival?*

A Surviving means staying alive. For animals in the wild, this means a daily battle against hunger and against being caught and eaten by other animals. It means finding water to drink, air to breathe, and shelter. And it means finding a mate so that new animals of the same kind will be born.

Q *What is the difference between instinct and intelligence?*

A Instinct is something an animal is born with. Intelligence is the capacity to learn, adapt to varying conditions, and remember.

All animals are born with certain instinctive ("inborn" or innate) behavior programmed into their genes. For example, all mammal babies know how to find the mother's nipple and suck from it.

Baby birds and some reptiles instinctively peck their way out of their shells when they are physically mature enough.

A baby chick in a barnyard instinctively pecks at anything on the ground that is of contrasting color to the earth: pebbles, grain, or sometimes its own feet! But it *learns* the best things to peck by its own experience and by watching its mother and other more experienced birds.

Q *Is it true that some animals, particularly insects and fish, look almost identical to their surroundings?*

A Yes. You may have seen—or tried to see—a leaf insect or a walking stick insect, perhaps in an exhibit at the zoo. It's almost impossible to spot these amazing creatures unless they move!

The walking leaf (of southeastern Asia) has veinlike markings, just like a leaf, and even sways gently, like a leaf in the breeze.

EXAMPLES OF CAMOUFLAGE

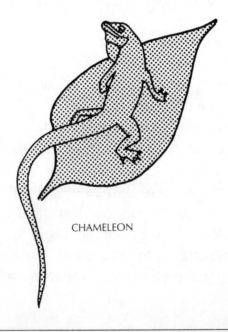

CHAMELEON

Another fake leaf is the leaf fish of South America. It lives among fallen leaves in streams. Its flat body seems to move exactly like a floating leaf.

A stick insect is long and thin—and it stays as still as a stick for hours on end.

Many moths can remain almost invisible, resting on a tree trunk that is the exact color of the insect.

Some animals that live in polar regions, where there is lots of snow, have fur or feathers that change with the seasons. The Arctic fox, for example, has rich brown fur during the short summer season. Its coat gradually changes to white as winter approaches. The ermine (a kind of weasel), the snowshoe hare, and the ptarmigan (a bird) are other examples of animals that change to winter white to blend with their surroundings.

SNOWSHOE HARE

Q *If disguise is so important for animal survival, why do some animals, such as zebras and tigers, have such vivid markings?*

A Strangely enough, because the markings make the animals hard to see when they are in their natural surroundings. This sounds strange, because vividly marked creatures look dramatic in zoos, which is where you usually see them. But it's amazing how the spots and stripes blend in with the light and shade of trees and grasses. The dark and light patterns also help to break up the outline of the animal, making it hard for the predator to tell which end is the head and which the tail.

Q *Do animals like to fight one another?*

A No. On the contrary, they may go to great lengths to avoid a fight.

Perhaps you have watched a cat or a dog when a stranger animal wanders into its territory (the place where it lives). The "owner" animal makes warning sounds (growls, hisses, or barks), and the hair on its body, but especially on the back of the neck, becomes erect. "Bristling" makes the animal look large and scary to the intruder.

The intruder may growl and bristle also, but now it thinks twice about making another step. After a moment of glaring and growling, it may sit down and start to wash its face or scratch its ears. It may even roll over on its back, which is the animal's way of saying, "Look! I'm harmless. You could grab my throat now. But let's be peaceful instead." And then it walks away. A good show was made by both sides, but nobody got hurt.

Q Do animals warn each other of danger?

A It certainly seems that they do, when we look at them and translate their gestures into human terms. (The scientific word for such a translation is *anthropomorphism*—pronounced an-thro-po-*more*-fism—which means roughly, "changing into man-form.")

For example, when a white-tailed deer flicks up its tail, all the other deer see the "signal." They take fright and run. But the action of the deer's tail may be just a reflex action (like the jerk of your knee when the doctor taps it with a rubber hammer) to a strange sight, sound, or smell. And the other deers' reaction to the flick of white also may be a reflex action. "White means run!"

Similarly, beavers sensing danger slap their paddlelike tails on the water. Hearing the signal, other beavers dive for safety.

Prairie dogs (small rodents related to squirrels) live in "towns," which consist of underground tunnels. While most of the animals go about their business of gathering and storing food and looking after their young, some prairie dogs stand upright, looking around them. When they bark out a "danger signal," all the others scurry to safety in their burrows.

Q *Do animals make homes to live in?*

A No, not the way people do. Birds build nests in which to lay their eggs, but once the babies have hatched out, the nest is abandoned, at least until the next breeding period.

Many animals seek shelter from rain, cold, and heat. They hide in caves and burrows, in crevices, under leaves, or in the mud. Animals may rear their young in these shelters, but they don't necessarily stay there once the babies can take care of themselves. In fact, many animals, such as deer, move their babies to different shelters every day.

Animals have to keep on the move, looking for food, looking for mates, and escaping from predators.

Some animals, however, make longer-term shelters. Among them are the social insects, such as bees, wasps, ants, and termites, some ground squirrels ("prairie dogs"), and some birds that form nesting colonies (for example, the weaverbirds.) Beavers make the longest-lasting shelters of all.

Q *Which animals dam up streams, lakes, and rivers in order to make their "homes"?*

A Beavers. These large rodents build the most astonishing "homes" in the animal kingdom. The beavers cut down trees with their sharp teeth. They carry logs and twigs, or roll them down the stream bank. Gradually, with mud and silt from the river, they create a wall, or dam, that holds back water. Then they start making a lodge. A beaver lodge has several compartments and underwater passages. Many generations of beavers live in the lodge and keep adding to it. When trees become scarce, the beavers move on.

Q *Do animals talk to each other?*

A No, not the way we do. But they do communicate with each other. Animals that live together in family groups—such as lions, wolves, whales, elephants, apes, monkeys, and social insects such as bees—have a most elaborate system of communication. It includes "body language," such as the twitching of ears or tails, scent markings, sounds, and (especially in the case of apes and monkeys) facial expressions.

Q *How do young animals learn to communicate?*

A In almost the same way that you do. "Social" animals—such as the ones mentioned above—watch the adults carefully and learn very quickly who is the leader of the pack, who gets the first choice at mealtimes, when to run and hide from enemies, how to interpret scent, sound signals, and body language. For animals in the wild, fast learning is a matter of life and death.

Q *Can apes communicate with humans?*

A We don't know yet. For years, scientists have been experimenting with different methods of communication. In the 1960's a chimpanzee named Washoe learned to express himself in the sign language used by the deaf. Other apes learned computer-based "picture" language. But the debate continues as to whether the clever apes are really "learning" and understanding language, or whether they are merely using repetitive behavior in the hopes of a reward or the fear of punishment. Keep checking!

Q *How do animals that live most of their lives alone, instead of in groups, find a mate?*

A In many different ways. Scent is the most common signal sent out, usually by the female when it is ready to mate. One particular scent, called a pheromone (pronounced FEE-ro-mone), attracts males from miles around.

Usually it is the male who puts on a big display. For example, the male frigate bird inflates its brilliant red throat to attract female birds flying overhead.

Some birds dance their way through courtship. Cranes will flutter their wings, stretch out their necks, stamp their feet, and bow to each other.

And one look at the peacock's gorgeous tail is almost guaranteed to impress the modestly colored peahen.

When a male fiddler crab is ready to mate, he digs a burrow in the sand. Then he stands at the entrance, waving one enormous claw. No female fiddler crab is likely to resist such a display.

Q *Why are crickets and katydids so noisy?*

A They are simply "talking" to each other. Except for the social insects, such as bees, wasps, ants, and termites that live together in huge colonies, most insects lead lonely lives. In order to announce their presence and find mates, they must send out signals, such as smells, lights, and sounds.

Cricket sounds are not made by vocal chords, but by rubbing together two hard, ridged parts of the wing tips or "knees."

Q *What is migration?*

A Migration is the periodic movement of animal populations away from and back to their usual environments.

A single round trip may take the entire lifetime of an individual, such as the Pacific salmon; or an individual may make the same trip over and over, as do many migratory birds and mammals.

Usually, the animals travel in groups along ancient, established paths (or flyways, in the case of birds).

Q *How do migrating animals know the way back "home," even though there may be no experienced adults to guide them?*

A The "homing" instinct of animals is another unsolved mystery, but here are a few clues:

We know that most animals have senses—such as hearing, smell, and eyesight—that are different, and in many ways superior, to those of humans.

For example, many animals can detect light and sound waves that are beyond the range of human capabilities. Migrating animals may hear the sound of a familiar waterfall or see reflected light from a lake, and use these clues as sign posts.

Another interesting clue: Recent studies have shown that many animals, including butterflies, dolphins, and bees (all of them migrate), have tiny magnetic particles in their brains. Scientists speculate that the magnetic particles act like compasses to give animals their built-in sense of direction. (People use man-made compasses that point to the magnetic north, and from there they can figure out all the other directions: south, west, and east).

So far, however, nobody has proved that animals use magnetic particles in their brains to figure out directions.

Keep checking!

Q *Why do animals migrate?*

A Nobody knows all the answers, but one reason for migration must be availability of food at any season; another reason must be to seek favorable grounds for mating and for rearing the young.

Q *Which land birds (as opposed to ocean birds that spend most of their lives flying over the ocean) hold the record for non-stop migratory flight?*

A As of 1980, the record was held by ducks and other shorebirds that flew the 48-to-72-hour journey, 2,500 miles (4,000 kilometers) from Alaska to Hawaii.

The birds gather in flocks in Alaska. A brisk wind from the north seems to be their signal for setting off, as, of course, it reduces their efforts by blowing them toward the south, sometimes doubling their normal speed.

Q *Do people migrate?*

A Yes, some do. In the Kalahari desert of South Africa some Bushmen still live as their ancestors did in the Stone Age. The women dig for roots and gather fruits and eggs. The men hunt for wild animals, such as antelope. When the area is hunted out, they move on.

In northern Africa, Berber tribes ride camels to herd their sheep and goats. They may settle for only a few days at a time, for there is not much grass in the desert. They are *nomads* (people who travel from place to place), and carry portable tents made from goat skin.

Some Eskimos, in northern Canada and Alaska, follow herds of caribou, musk-oxen, and also hares and foxes. They hunt seals, walruses, and polar bears. Often on the move, these Eskimos may build temporary shelters from blocks of ice. These shelters are called igloos. But most Eskimos today have never seen an igloo. They live in permanent settlements and towns.

Q *Is it true that eels from Europe and from North America meet at one place in the mid-Atlantic to breed?*

A Yes. It is strange, but true. Eels are long, slithery fish that look a little like snakes. They spend most of their lives in freshwater lakes in North America and in Europe.

When breeding-time comes, eels from both continents travel all the way to the Sargasso Sea, an area in the mid-Atlantic between Europe and North America. There, the eels mate with those of their own species. The young eels, called elvers, travel back to either Europe or North America. Somehow, they never get mixed up or go in the wrong direction.

Nobody knows why eels make this strange and difficult journey in order to mate.

Q *What is the "food chain"?*

A The food chain is a phrase made up by scientists to express the fact that each animal is dependent on another animal, who in turn is dependent upon plants, in order to stay alive.

All animals must eat to live. Animals that eat meat are called *carnivores* (most long-toothed hunters such as cats and wolves, for example). Animals that eat plants are

herbivores (most cattle, for example). There are also *insectivores* (such as anteaters, who eat insects), and *omnivores* (such as people,) who eat plants and animals.

Q *Is it true that some birds lay their eggs in nests made by other birds? And that the "other birds" bring up the "stranger" birds?*

A Yes. Some birds, such as the European cuckoo and the North American cowbird, lay their eggs in nests of smaller birds.

When the young cuckoos or cowbirds hatch from their eggs, they are big enough to push the other birds out of their nest. But the parent birds respond to the food-demanding behavior of the strange, big birds, and keep on feeding them, as if they were their own "children."

Scientists haven't figured out why some birds behave in this manner, except to say that it's just another example of the way that animals behave in order to survive.

Q *What is a good example of a food chain?*

A A freshwater pond. Plants grow in and around the shallow waters of a pond. Hundreds of insects and other animals feed upon the plants and take shelter in their leaves.

Insect and fish larvae feed upon the parts of the plant that grow underwater. Larvae also feed upon each other, and are in turn eaten by larger animals.

Other animals feed on the parts of the plant that grow above water. Animals also help to spread the seeds from one plant to another of the same kind.

Birds, fish, turtles, and frogs feed upon the insects that live on the pond plants.

Birds, such as ducks, eat the pond plants. Other birds, such as herons, feed upon fish and frogs who live among the pond plants.

Otters and other animals who live around the pond feed upon fish, frogs, eggs, and insects.

A dead animal sinks to the bottom of the pond. Its body is eaten by fish, the larvae of fish and insects, and tiny, microscopic bacteria. Its body becomes part of the nourishing water and soil of the pond.

A freshwater pond is only one example of a food chain, or *ecosystem*. Grasslands, deserts, mountaintops, seashores, coral reefs, woodlands, all have their own ecosystems.

FOOD CHAIN OF FRESHWATER POND

PLANTS

DEAD ANIMAL

HERON

OTTER

FISH

BIRDS

FROG

DUCK

INSECTS

FISH LARVAE

PLANTS

Each animal depends upon another for its food, and all animals, in the end, depend upon plants for survival, for only plants are capable of manufacturing their own food from sunlight and air.

Q *Is it true that whale babies have "aunts"?*

A Yes. A whale is a mammal that lives in the ocean. Mammals have to breathe air (just as people do). Whale babies are born in the ocean. The mother whale comes as close to the surface of the ocean as she can, when she gives birth to her baby calf. Once the baby whale is born, a nearby "aunt," or female whale other than its mother, may help by pushing the baby up to the surface so that it can draw its first breath of air.

Q *What other large animals have "aunts" or other adults that look after them?*

A Elephants are among the other large mammals that seem to "look after" the offspring of others in the same family group.

When a baby elephant is born, the other elephants crowd around it, protecting it from lions and other predators.

Q *Which reptile not only takes care of its eggs, but looks after the babies when they are born?*

A The crocodile. The large and toothy mother stays near the eggs that she has laid in a mound of vegetation or buried in the sand near the water. She drives off intruders.

When the babies hatch out, she may carry some of them in her mouth down to the water. She stays with the babies for several days, sometimes letting them ride on her back.

Alligators, closely related to crocodiles, show the same behavior toward their young.

Q *Do fish make nests for their eggs?*

A Most of them do not. They lay millions of eggs that float freely in the sea. But some fish, such as the male stickleback, make nests and stay close until the youngsters hatch and swim away.

Q *Do frogs and toads look after their eggs?*

A Most of them do not, but there are some startling excep-
tions. The father Rhinoderma, a tiny frog from southern
Chile, appears to eat the eggs that have been laid upon
moist ground. But he doesn't swallow the eggs. He keeps
them in his exceptionally large throat sac. There the eggs
safely develop and hop out of the father's mouth as fully
formed froglets.

Another unusual parent is the Pipa toad, also South
American. It lives entirely in water. But instead of leaving
its fertilized eggs to float freely, the male toad carefully
scoops them onto the female's back. The eggs stick there
and gradually skin grows over them. After about two
weeks, the female's back begins to ripple with the movement of the developing tadpoles. Then, after another 24
days or so, the tadpoles break through the protective skin
and swim away.

A European species, the male Midwife toad, scoops up
the strings of eggs and twines them around his legs. He
hobbles around with them for a few weeks, staying in
moist places. When the tadpoles are ready to come out of
the eggs, the father dips his legs in the water and waits for
all the babies to emerge.

Q *Is it true that some birds put on an act to lead predators away from the nest?*

A Yes. Just as one example, the Golden Plover runs from its nest, calling loudly and dragging its wings as if they were injured. The predator naturally goes after what looks like an easy catch. The plover lets the enemy get close, then it suddenly takes off in full flight right under the hunter's nose. Because the animal distracts the attention of the hunter, scientists call this act a "distraction display."

Q *Can animals change their colors?*

A Yes. If an animal such as a flounder or a chameleon stays long enough in one place—a particular patch of sand and gravel on the bottom of the sea for the flounder, or a leafy branch for the chameleon—its pigment (color) cells gradually activate more or less color until the animal becomes the same color as its surroundings.

Q *Birds have no teeth. So what is the "egg tooth" that a baby bird uses to break out of its shell?*

A It isn't a real tooth. It is a horny growth on the tip of the beak. It falls off a few days after the baby has hatched.

Q *Does the phrase, "eating like a bird," really mean eating small quantities of food?*

A Yes. The phrase is used to describe a small eater, but it's a most inaccurate term. Birds eat all day long and sometimes at night. A hummingbird, whose length is no longer than your little finger, eats many times its weight in food every day. It needs food to create the enormous energy that powers its wings. A hummingbird's wings beat 50 to 75 times a second. Other fast-flying birds, such as swifts and swallows, fly with their mouths open in order to catch flying insects. Swifts and swallows are in flight for most of their lives.

Q *How do birds know that they must feed their young?*

A Parent birds react instinctively to the sight of the open mouths of their young. Baby birds open their mouths wide and make "demanding" (peeping) sounds. Inborn behavior patterns are triggered by the sight and sound combinations to make parent birds feed their babies.

Q *Why do birds sit on their eggs?*

A Eggs must be kept at just the right temperature so that the growing chicks can mature. Sometimes this means that the parent birds must keep the eggs warm with the heat from their bodies.

In warmer climates, parent birds must try to keep their eggs from overheating. The killdeer *(Charadrius vociferus),* for example, which lays its eggs on rocky ground where the hot sun could bake the eggs, often shades them with outstretched wings. Sometimes the killdeer soaks its belly in water from a nearby stream and then lowers itself over the eggs to cool them with water.

Father penguins, in the freezing Antarctic, hold the eggs on their feet and cover the eggs with a flap of flesh that hangs from the abdomen. (The mother penguin is off feeding at this time, and returns when the babies hatch.)

Mallee fowl, of Australia, bury their eggs in large mounds of earth and leaves. The male parent keeps testing the temperature of the mound with his beak (probably with his tongue). He rearranges the earth and leaves to adjust the temperature, which is generated by decaying vegetation, heat from the earth, and heat from the sun.

There are almost as many kinds of nests as there are birds.

Q *How do most birds feed their newly hatched babies?*

A The parent birds swallow food, digest it, then bring up (regurgitate) a kind of baby puree that they place in the open mouths of their young.

Q *Why do rattlesnakes, and other snakes that kill with poison, allow their victims to run away after having been bitten?*

A Because, although a snake's venom is powerful, the snake itself is a fragile creature. A struggling rodent, such as a rat, could cause dangerous wounds on the snake's delicate body.

The snake lets its prey run off, but its instinct tells it to follow at a safe distance. The stricken prey is usually overcome by the poison and dies in a matter of minutes, at which time the snake can dine at leisure—and in safety.

Q *Do pigeons live only in cities?*

A No. Pigeons (or rock doves) also inhabit the countryside. Long before there were any cities, these birds were nesting on rocky ledges and in crannies. Building exteriors supply thousands of nesting places for pigeons, and people supply food, so it's not surprising that these birds are found in cities all over the world.

Q *Why are earthworms useful to gardeners?*

A Because they help keep the soil loose and full of tiny air pockets. Earthworms live mostly underground, hollowing out little tunnels as they move. Worms also eat decaying plant matter and sometimes small, dead animals. This food is ground up by the worm's digestive system and excreted as fine "casts." The worm's tunnels and casts make good soil for growing plants.

Q *Is it true that a worm has no eyes or ears?*

A Yes. In human terms, a worm is blind and deaf. However, all of its surfaces are sensitive to light and to vibrations. People who use worms as fishing bait take advantage of this fact by shining a flashlight on the ground at night and wriggling a stick in the ground. If there are worms around, they are disturbed and come squirming up to the surface.

Q *Is it true that if a worm loses part of its head or tail, that part will grow back again?*

A Yes. If a worm is in danger of being caught by a predator, it can cast off a part of its body. This reflex action is called autotomy ("self-cutting"). Many lizards, starfish, and other animals have this same method of self-preservation. The remaining portion can often regrow the lost part or parts, but the new growth is limited in extent, and is never "as good as new."

Q *It is a well-known fact that "social" insects, such as certain bees, wasps, ants, and termites, take care of their young. Are there any other insects that look after their young?*

A Yes. The female earwig *(Forficula auricularia)* lays fifty to one hundred eggs in a hole, then shuts up the entrance and guards it. She constantly licks the eggs to keep them warm, wet, and clean. If danger threatens, she may carry the eggs to another hole. Even after the eggs are hatched, the mother keeps an eye on them. The youngsters are tiny, but grow larger with each molt (skin shedding). If they feel endangered, they may return to their original nesting hole for shelter, at least until they are full-grown.

There are about nine hundred kinds of earwig, ranging from the common ones in your backyard (about ½ inch or 1.25 centimeters long) to tropical species of about 2 inches (5 centimeters) in length.

Earwigs do *not* climb into people's ears. They much prefer flower petals or cracks in stony walls as shelters.

Q *What causes a phosphorescent (fos-fuh-RES-cent) glow in sea water after dark?*

A The light is produced by microscopic, one-celled creatures, which are part animal and part plant. They are generally known as flagellates. Millions of them are present in ocean waters. These minuscule creatures respond to disturbance—such as the movement of a night swimmer or a boat—by flashing on the same kind of light that is produced by fireflies: a cold, rather than a warm light.

Nobody knows why or how these tiny specks in the ocean produce light. Perhaps it is their way of showing that they are irritated or disturbed.

Q *Is an octopus and a squid the same kind of creature?*

A These two are closely related and share many of the same habits. The main difference is that an octopus always has eight arms, or tentacles, while a squid has ten.

Both of these animals belong to the same mollusk family as clams, oysters, the nautilus, and the cuttlefish. But in many ways they are different and more "intelligent." For one thing, both octopi and squid have eyes that are very much like human eyes. They perceive

things in much the same way. (If you have been swimming with a group of squid in a coral reef, you will notice how they seem to watch you, and move backward and forward at the same rate that you do.) Squid can grow to the enormous size of 66 feet (20 meters) long and weigh as much as 42 tons (38,000 kilograms). And yet, both octopi and squid are gentle, shy creatures who seem to prefer fleeing to fighting.

Q *Do "flying fish" really fly?*

A No, not in the way that birds and insects do. But a flying fish can move fast enough near the surface of the water to gather speed and shoot above the water at a great rate, especially if a hungry tuna is chasing it. Once above water, a good wind may lift the fish on air currents and it may reach a height of 10 to 20 feet (3 to 6 meters) before starting to fall back into the water. Flying fish have strongly developed pectoral fins and tails. Even on windless days they can taxi along the surface of the water, vibrating their tails, and moving along for at least 50 yards (45 meters) before descending back into the water.

The California flying fish (*Cypselurus californicus*) is the largest of its kind: it grows to about 18 inches (45 centimeters) long.

Q *Why do shore birds tuck one leg up while they rest, and is it always the same leg?*

A The birds stand on one leg to give the other leg a chance to rest and keep warm. They switch legs from time to time.

Q *What fish can live on dry land for several years, with hardly any air and no outside source of food or water?*

A The African lungfish *(Protopterus annectans)*. This remarkable creature can bury itself in the mud and wait out a dry period for as long as four years until the next rainfall.

Q *How does the lungfish survive on land?*

A It creates a leathery, mucus-lined cocoon around itself and literally feeds upon itself. It doesn't use its own fat, the way some hibernating animals do. Instead, it absorbs its own muscle tissue. Its remarkable kidney filters water

from the fish's own toxic wastes and keeps recycling it. When the lungfish finally comes out of its little "space capsule," it is smaller and lighter, and stiffly curled up into a ball—but still alive. No wonder this little creature has survived for about 390 million years, with or without water!

THE HUMAN BODY

It's amazing how little we know about our bodies, considering that we live in them all of our lives!

For example, do you know how big your heart is, or where your kidneys are? Do you know why blood is red? Why we yawn, shiver, get goosebumps? Why it doesn't hurt to cut nails or hair? Do you know how babies start to grow and what twins are?

You'll find the answers to these and many more questions in the next few pages about that amazing, intricate, highly efficient collection of organisms called the human body.

Q Why is the human brain compared to a computer?

A Because, like the computer that it invented, the human brain is able to receive, sort out, store, and send out millions of pieces of information.

The entire nervous system of the body keeps sending messages to the brain all day long—messages like "Breathe!" "Swallow!" "Step back!" "Cough!"

All the nerves in the body are connected to the brain by the Central Nervous System. Many nerves act automatically with the brain to keep you alive. For example, you don't have to tell your heart to beat or your lungs to breathe or your digestive system to digest. The brain is automatically "programmed" to send appropriate messages to various organs of the body.

Some parts of your brain act because "you," that is, the non-automatic part of you, tell them to. For example, you may have developed good habits or bad habits. You have programmed your brain and your body to act in a certain way.

One of the most interesting things about the human brain versus the computer is this: the human brain can change its "mind" or "plan" very rapidly. The computer has to have new information fed into it, and old instructions erased, in order to act in a new way. Thus, it cannot "think" for itself.

Q *Which organ "gives orders" to the rest of the body?*

A The brain. Protected by a bony skull, the brain looks a little like a gray, pulpy cauliflower. It is made up of millions of cells. Nerves from all over the body connect with the brain through the spinal cord and the neck.

Different parts of the brain receive, sort out, store, and otherwise respond to different parts of the body, particularly to the sense organs: the eyes, the nose, the tongue, the ears, the skin.

It is the brain that sends messages to all your organs, so that, automatically, the heart beats, the lungs breathe, the digestive system digests.

The body can get along with only one kidney and with an artificial heart and limbs, but without the brain no other part of the body could function.

Q *What is the mind?*

A It is the part of the brain that remembers and helps you to create stories and music and dreams; it is the part of the brain that thinks and helps you to solve problems. The right side of the brain is the creative half; the left side is the practical half that can figure out complicated problems such as mathematics and language.

Strangely enough, the left side of the brain controls the right side of the body. The right side controls the left side of the body.

Q *How big is your heart?*

A It is about the size of your fist. As you grow bigger, so will your heart.

Q *How fast does the heart beat?*

A It depends upon how big you are. In general, the larger the person, the slower the heartbeat. For example, a man's heart beats about 70 times a minute, a woman's a little faster, and a baby's about 130 times a minute. (An elephant's very large heart beats about 25 times a minute.)

Q *Why is it necessary for the heart to keep beating for you to stay alive?*

A Because the heart is the muscle that pumps life-giving blood throughout the body, all day, every day, from the time a baby is in its mother's womb until the time a person dies.

Q *What is the difference between a vein and an artery?*

A An artery carries blood away from the heart. A vein carries blood toward the heart.

The blood in the artery contains dissolved fresh oxygen which has entered the heart chambers, by way of major blood vessels, from tiny air sacs in the lungs.

The blood in the veins contains waste material which it carries up to the heart. The waste gases from the veins filter through to the air sacs in the lungs and are breathed out when you exhale.

Q *Why is it harmful to hold your breath for more than a minute or so?*

A It is harmful because your body isn't receiving its normal supply of oxygen, nor is it able to get rid of waste gases that can poison your bloodstream. When you hold your breath, two things happen: the supply of fresh oxygen is cut off from your lungs and, therefore, from the rest of the body; and the lungs fill up with waste gases (mainly carbon dioxide) that escape when you breathe normally.

Q *How much air do you take in with each breath?*

A About one pint when you are resting (shallow breathing) and as much as four quarts when you are taking deep breaths or running or exercising.

Q *Why do we yawn?*

A To take in more air. Often we yawn when we feel sleepy. The body is relaxed and the breathing is shallow. Opening the mouth wide and taking in extra air in a yawn is a reflex action. It is something you cannot control (though you may be able to stifle it, with great effort). The extra oxygen speeds up the circulation and stretches the muscles. Perhaps, long ago, when our ancestors had to stay alert out in the wild, they yawned to keep themselves from falling asleep when there might be danger.

Sometimes we yawn when we are bored, which is almost the same as being sleepy: the senses are not being stimulated, we relax, and our breathing becomes shallow. The lungs demand more air, and our reflex action is to yawn.

Q *How much blood is in the human body?*

A It depends on how big you are . . . and on where you live. If you weigh about 80 pounds (36 kilograms), you have about 2½ quarts (2⅓ liters) of blood in your body. Someone double your weight has twice as much blood.

At high elevations, such as at mountaintops, there is less oxygen in the air than at ground level. Your body must manufacture more blood to capture and circulate the extra oxygen that the body needs to function and develop.

Q *How long would all the blood vessels in your body be if they could be straightened out and measured?*

A About 100,000 miles (161,000 kilometers) long. That is, long enough to circle the world at the equator *four* times, or as long as all the miles of railroad tracks in the United States!

Q *How can the human body hold so many miles of blood vessels?*

A Because most of the blood vessels are thinner than a hair's breadth. The largest blood vessels are the arteries and the veins. These are like the great superhighways of the body. But there are also millions of tiny *capillaries* which are the tiny roads, paths, and trails of the blood system, unmarked on most maps, and, in the body, invisible except through a microscope.

Q *How long does it take for blood to make its round-trip journey to and from the heart?*

A Less than a minute! The pumping heart muscle loses no time—it never stops. In a healthy body there are no traffic jams, since arteries and veins are one-way streets. Thanks to very efficient valves (which act like trapdoors), the blood can flow only one way, make a quick side trip to the lungs for an exchange of air, and then start on its way again.

Q *Why is blood red?*

A Because the bloodstream contains more red cells than any other cells. Besides the red cells, blood consists of white cells (about one for every thousand red cells), plasma, a watery liquid in which the blood cells float, and platelets, which are cells that cause the blood at a wound to clot and so prevent loss of blood.

COMPONENTS OF BLOOD

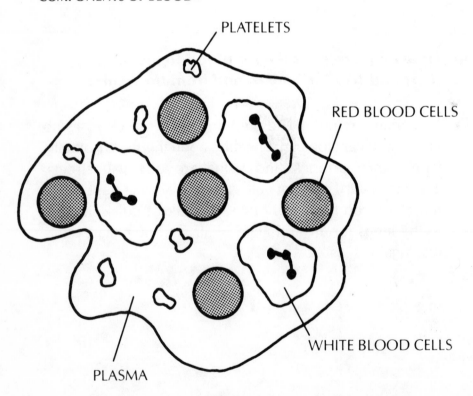

PLATELETS

RED BLOOD CELLS

WHITE BLOOD CELLS

PLASMA

Q *What is a bruise?*

A A bruise is an injury in which the skin is not broken but is discolored when blood vessels are ruptured and blood leaks into tissues under the top layer of skin.

Q *Are bruises always "black and blue"?*

A No. A bruise is usually greenish yellow. When the injury occurs, hemoglobin and other chemicals in the blood break down into their various components. Iron from hemoglobin gives the bruise a greenish tint and other chemicals cause variations of blue and green and purple.

Q *How does a vaccine work to prevent illness?*

A A vaccine works by stimulating the body to produce *antibodies*, the chemical "soldiers" that fight off invading bacteria.

Vaccines are grown in the laboratory from diseased cells.

A tiny amount of the vaccines injected into the body gets the antibodies all worked up and ready for a fight. If an illness actually strikes, the body is already mobilized to destroy the invaders before they can do any damage.

Q *What is "hay fever"?*

A Hay fever is a common and inaccurate term for an *allergic* reaction to substances such as pollen, animal hair, and certain drugs and foods. An allergic reaction isn't usually triggered by hay, and there is no fever.

Q *What is an allergy?*

A An allergy is an overreaction of the antibodies in our system which protect us from foreign substances. Practically everybody sneezes or coughs when dust is in the air.

Sneezing and coughing are normal reactions: the body is trying to get rid of dust that has been taken in through the nose and mouth. But with some people, the body's "soldier" cells—the antibodies—get "hysterical" and cause skin rashes, swelling, coughing, weeping, sneezing, to such a degree that a person could be endangered by not being able to breathe.

Q *Is there any way to get rid of an allergy?*

A No. Usually your body is stuck with those vigorous antibodies that take offense at allergens (substances that cause the violent reaction).

However, there is some help. Certain drugs called *antihistamines* help to calm down the antibodies that produce the excitable histamines in your blood system. There are also decongestants that help clear up a stuffy nose or throat and runny eyes. In some cases, regular innoculations can help prevent allergic reaction. And, of course, staying away from the cause of the reaction always helps. Sometimes allergies disappear. Sometimes they suddenly appear. It's always best to consult a physician if an allergy is frequent and bothersome.

Q *What is a vitamin?*

A A vitamin is an essential nutrient required in the diet for the body to function properly. A person eating a varied and well-balanced diet gets all the vitamins he or she needs to make all the different food particles work well together to nourish the body.

Q *How is a piece of food changed from its solid form so that it can pass into the bloodstream and circulate throughout the body?*

A Food is changed into tiny liquid particles through the digestive process. This process starts even before we put food into the mouth. The sight and smell of the food starts the salivary glands working. Inside the mouth, the saliva mixes with the food to soften it, the teeth chew and grind the food, and the tongue helps to push the soft mass down the throat.

Food travels down to the stomach via the esophagus, a kind of pipe that lies in the chest.

The stomach is a sac that lies quite high up underneath the ribs (not underneath the navel, as most people think).

The stomach contains digestive juices that, along with the muscular churning of the stomach walls, further soften the food. After a few minutes or hours (depending on the kind of food), the food passes into the small intestine as a liquid.

Here the liquid is worked on by more juices. All the digestive juices, beginning with saliva, work to separate the various elements of which food is composed: carbohydrates, fat, protein, vitamins, minerals, glyceroids. When the food is completely liquid, it is absorbed by the walls of the small intestine and seeps into the bloodstream.

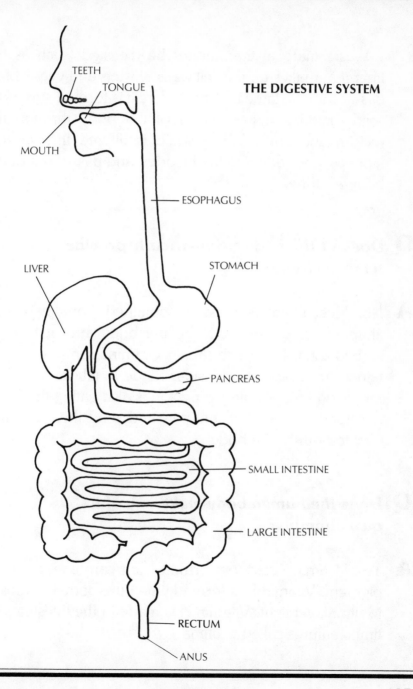

THE DIGESTIVE SYSTEM

TEETH

TONGUE

MOUTH

ESOPHAGUS

LIVER

STOMACH

PANCREAS

SMALL INTESTINE

LARGE INTESTINE

RECTUM

ANUS

Waste material that cannot be digested (such as the fiber that makes up the cell walls of fruits and vegetables) passes into the large intestine. Here it is formed into more solid material. It passes out of the body through the rectum and anus. Waste liquid is filtered through the kidneys, goes down to the bladder, and passes out of the body as urine.

Q *Doesn't the body automatically provide its own vitamins?*

A No. Most vitamins must be obtained from plant and animal sources, or made by the body after it has absorbed a particular substance. Vitamins can also be obtained from pills and capsules made in laboratories, but they won't work unless a person is also eating the foods necessary to mix with the pills and send them on their way to nourish the body.

Q *Does the human body make any of its own vitamins?*

A Yes. Vitamin A can be formed from carotene, a yellow pigment. Vitamin D is formed when the body is exposed to ultraviolet light. Vitamin K is made in the intestines by tiny creatures called bacteria.

Q *Can something indigestible be good for you to eat?*

A Yes. The fiber found in plant foods (such as celery, strawberries, potatoes, cabbage) cannot be digested by the human body. The fiber from the plants bulks up in the intestines, giving a feeling of fullness, and also helps the intestines to work more easily.

Q *What is fiber?*

A Fiber is the cell wall that surrounds every cell that makes up a plant. Animal cells don't have it. Only plants have fiber.

Q *Why can we tolerate drinking hot liquids, while the same liquid would scald our skin?*

A For two reasons: as we sip a hot drink, we also sip in air, which cools the liquid; and the saliva in our mouths dilutes the liquid and cools it before we swallow it.

Q *Where is the liver and what does it do?*

A The liver takes up a large part of the right side of the abdomen, just above the waist.

The liver is a very busy organ. Like the kidneys, the liver filters out impurities from the blood. It stores fats and sugars and releases them as they are needed. And it produces digestive juices which help the small intestine to break up food and digest it.

Q *Where are the kidneys and what do they do?*

A The kidneys are two bean-shaped organs, 4 to 5 inches long, situated on either side of the spine a few inches above the waistline, that is, approximately in the middle of your upper back. (It's amazing how many people don't know where their kidneys are!)

The kidneys act as filters to purify the blood. They squeeze out waste materials and send them through the bladder as urine. The "clean" blood is then sent on its way through the bloodstream to nourish the body.

Q *What is the largest organ in the human body?*

A The skin. The total skin weight of an adult is about 6 pounds. Skin, and the hair that grows on it, covers every part of the body in a smooth, waterproof, snug-fitting coat. Although you can't always see them without a magnifying glass, there are tiny pores all over the skin. (An adult has about 2,000 pores per square inch!) These pores allow the body to get rid of wastes and excess heat through perspiration, and thus keep the body at exactly the right temperature at all times.

Q *Why do we shiver when we are cold or frightened?*

A We shiver when the tiny muscles near the surface of the skin contract (tighten) in response to cold. This reflex action of the muscles does two things: it raises the hair on the skin, and the muscular action creates some warmth.

The raised hair (which we can see sometimes only in the form of "goosebumps") helps trap warm air from the body. The raised hair on furry creatures (such as dogs and cats, and on our hairy ancestors) not only traps heat. It makes the animal look bigger and therefore more frightening to a potential enemy.

Q *What gives skin its color?*

A Skin gets its color from *melanocytes*. Melanocytes are cells which produce the brown color, *melanin*. These cells are produced in the lower level of the epidermis. Everybody has the same number of melanin cells. What causes a difference in skin color is the way that the cells are distributed in the skin and the quantity of melanin that is produced.

For example, since melanin helps protect skin against the dangerous, burning, ultraviolet rays from the sun, more melanin is produced by the cells when a person is exposed to sunlight. In a "fair-skinned" person there are only a few grains of melanin, but if that person lies in the sun, production in the melanin cells speeds up and the person begins to look darker in color—if he or she is lucky enough not to get a dangerous burn.

In a dark-skinned person, the melanocytes produce lots of melanin all the time. These people can stand a little more sunshine without getting burned, but not much. Sooner or later their blood vessels swell up and the skin becomes sore and burned.

Q *Do people who live in hot countries have darker skin than those who live in a cool climate?*

A In ancient times, when people didn't travel far, the answer would have been, "Yes." The melanin-producing cells of people living in constant sunshine produce more protective coloration against the dangerous ultraviolet rays of the sun. People in cool, northern climates need to produce less protective coloration.

Today, people travel and settle all over the world, and their skin color may have very little to do with where they live.

Q *Do people inherit skin color from their parents?*

A Yes. The color of your skin, hair, and eyes, is inherited from parents and other ancestors. But since a great many ancestors are involved, on both your mother's and your father's side, you are made up of so many thousands of genes that you are uniquely you—there is nobody in the world exactly like you.

Q *What are freckles?*

A Freckles are spots or blotches of dark brown on the skin. They are produced by the activity of melanocytes (color cells) in the outer layer of skin. For some reason, some clumps of melanocytes produce more melanin than others. Freckles are most easily seen in fair-skinned people and become more prominent when they are exposed to sunshine.

Q *Is it true that people have hair all over their bodies?*

A Yes, though many of the hairs are invisible without the use of a magnifying glass. Scientists think that our ancestors were as hairy as apes. As our species became more numerous, people began to travel to colder climates. They began to wear clothes (mostly animal hides) for protection against the cold.

   As thousands of years went by, the need for hair became less vital, except in especially sensitive areas: the head, the chests and backs of some men, the legs and arms, the armpits, and the pubic areas of adults of both sexes.

Q Do human beings shed their skin?

A Yes. Some skin cells are shed every day from the dead outer layer called the epidermis. New skin cells are constantly growing in the dermis and pushing their way outward. You can sometimes see dead skin in the form of flakes (especially after a sunburn) and in dandruff from the scalp.

SKIN

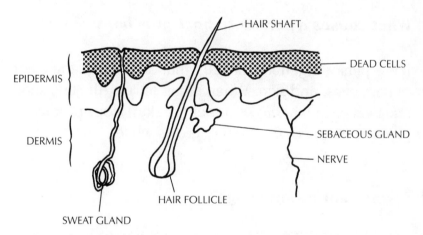

Q How long does it take to grow a new coat of skin?

A About 30 days . . . but since we don't shed our skins all in one piece, as snakes do, we do not see the renewal taking place, except where the skin has been injured.

Q *Why doesn't hair grow on the palms of the hands and the soles of the feet?*

A No one knows for sure, but one theory is that there is no advantage to having hair on areas used for grasping and walking. Another explanation may be that on those much-used parts of the body hair gradually wore away and the skin became tougher and thicker, giving more protection from rough surfaces.

Q *What causes a person to have gray hair?*

A It is a person's genetic makeup that determines the color of hair, eyes, and skin. When a hair follicle (the tiny shaft in which each hair grows under the skin) stops producing pigmentation (color), the hair grows white.

Q *Is gray hair a sign of age?*

A No, not necessarily, although hair follicles tend to produce less and less color as a person grows older. However, if a person's genes are "programmed" for that person to have gray hair at a certain age, even teenagers can have gray hair.

Q *Can a person's hair suddenly turn white from shock?*

A No. The loss of color is a gradual, natural process and has nothing to do with one's state of mind.

Q *Why doesn't it hurt to cut nails and hair?*

A It doesn't hurt because both nails and hair are essentially dead substances.

The growing part of a hair follicle or a nail is located in the dermis, beneath the outer layer of skin.

When the hair follicle reaches the surface of the skin, where you can see it, it is being pushed out by the active growing of hair follicles underneath. Threads of hair become longer and longer, but once they are visible on the surface of the skin they have no nerves or blood vessels to feel pain. That is why it is painless to cut hair or to shave.

Similarly with nails: the tip of the nail is dead. It appears to be growing because the cells from the lower part of the nail, underneath the cuticle, keep growing and pushing the dead cells onward. It doesn't hurt to cut nail tips because the nails have no nerves—and therefore no feeling.

Q *How many bones are in your body?*

A There are 206 bones in your body. Bones form your skeleton. The tiniest bones, the ossicles, are inside the ear. The largest bone is the thigh bone, or femur (in your upper leg). More than half of the bones consist of the tiny bones in your hands and feet.

Q *What is inside a bone?*

A Inside the bone is a spongy tissue called marrow. Red and white blood cells are made and stored inside the marrow and released into the body as they are needed.

Q *Is it true that a baby's bones are not fully formed?*

A Yes. Many of its bones are still in the form of *cartilage,* a flexible, gristly tissue, such as that which makes up your nose, ears, and part of your ribs. Everybody's skeleton (bony framework) starts out as cartilage and hardens gradually. Even a baby's skull bones are soft and not yet joined together. This flexibility makes it easier for the baby's head to come through the mother's birth canal. By the time a child is twelve, the bones are hard and fully formed although still growing in length.

Q *Bones seem so stiff and solid. How are we able to bend our arms and legs, hands and feet, neck and spine?*

A It is joints that make our bodies flexible. Joints are places where the bones of the skeleton fit together. The sockets of the joints are lined with a fluid that makes movement easy in most parts of the body. Elbows and knees are examples of movable joints.

Q *Are there any movable joints in the skull?*

A Only one: the lower jaw. Its movement allows us to open and close our mouths and to chew our food.

Q *What does being double-jointed mean?*

A It means having an unusual amount of flexibility at certain joints—often at the elbows, knees, shoulders, and fingers.

The term itself is nonsense, because everyone has the same number of joints.

Children whose bones have not yet completely formed—that is, those who still have flexible cartilage near their joints—often can move their joints in ways that become impossible later on.

Q *Does the length of your thigh bones determine your height?*

A Yes—along with other inherited factors. Take a look at a row of people sitting down. If they are all adults of the same sex, they all look about the same height. But once they stand up, you can clearly see that some people are a few inches taller than others! That's because the length of their torsos (neck to top of leg) are all about the same—about 28 inches for a man, and 24 inches for a woman. Once they stand up, the length of their thighs makes the difference in height. The femur, or thigh bone, makes up $2/7$ of a person's total height.

Q *What is the funny bone?*

A It is the name sometimes given to the bone in your upper arm. Its correct name is the humerus (from the Greek word for shoulder, which is the upper joint of the humerus). The humerus connects with the bones of the forearm (the radius and the ulna) at the elbow, where there are some particularly sensitive nerves near the surface. If you hit your elbow against something hard, you get a strong, tingling sensation. It isn't funny at all: it can be quite unpleasant for a moment. But long ago some jester made the "connection" between the word humerus and humorous ("funny"), and so the words funny bone, became part of the language.

Q *How do babies start to grow?*

A Babies start to grow from two cells: a female ovum (egg) and a male sperm. The two cells join together inside the mother's body. Then the cells start dividing and multiplying and become a tiny human embryo. The embryo grows inside the mother's womb, or uterus, where it is provided with oxygen and nutrients (dissolved food) for about nine months, or until it is ready to be born.

Q *How long does a person keep growing?*

A The human body keeps growing in height and weight for about 18 to 20 years after birth, but not at the same rate year after year.

A baby that weighed 7 pounds (3 kilograms) at birth may weigh three times as much (about 21 pounds) (9½ kilograms) when it is one year old. At age six, the child will have gained another 20 pounds (9 kilograms) or so. But then growth slows down—otherwise, the child would be about 20 feet (7 meters) tall and weigh 7,000 pounds (3,185 kilograms) at age ten!

Q *What are you made of?*

A Cells: thousands and thousands of cells so tiny that you can see them only with a powerful microscope. Cells come in different shapes and sizes: there are bone cells, muscle cells, brain cells, blood cells, nerve cells, and many more.

Each cell has a special job to perform. For example, blood cells, which are the only ones to move freely all over the body, carry dissolved air and nutrients to all the other cells.

MUSCLE CELLS

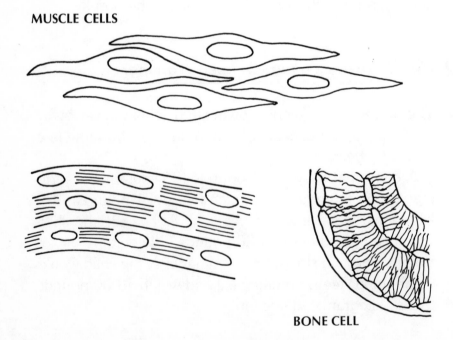

BONE CELL

Q *Do boys and girls grow at the same rate?*

A No, not all the time. They grow at about the same rate until they are about ten to twelve years old. Then girls start to develop much quicker than boys, for a few years.

But we all grow at different rates.

Boys usually start catching up with girls when they are about thirteen to fifteen years old. They usually grow taller and heavier than girls at this age.

Q *What is heredity?*

A Heredity is what you inherit from your parents, grandparents, and other ancestors. It is the code that is programmed into the mother-and-father genes that joined inside your mother's body when fertilization took place.

Heredity determines the color of your skin, hair, and eyes, and also many other factors, such as the shape of your nose, whether you'll have straight hair or curls, long legs or short legs.

Q *What decides whether you'll be a boy or a girl?*

A It is the chromosomes in the male sperm that determine the sex of the child.

Chromosomes are the microscopic particles that lie at the center, or nucleus, of a cell. The chromosomes contain the genetic code that determines many characteristics, including sex.

Each cell contains twenty-three chromosomes. In a female, the sex-determining chromosomes are always X, or female cells. The male has both Xs and Ys. If two X chromosomes unite during fertilization, the resulting child will be female. If an X and a Y unite, the child will be male.

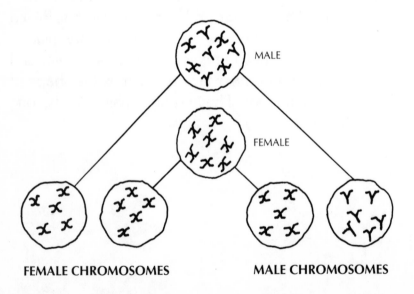

MALE

FEMALE

FEMALE CHROMOSOMES   **MALE CHROMOSOMES**

Q *What are twins?*

A Twins are two babies that grow in the mother's womb at the same time and are born almost at the same time. (One comes out usually just minutes ahead of the other one.)

Q *What causes twinning?*

A Twinning is caused when the fertilized egg in the mother's womb splits in two identical halves. (Usually the fertilized egg remains as one unit which develops into one baby.)

Q *What is the difference between identical twins and fraternal twins?*

A Identical twins are formed when one female egg (ovum) is fertilized by one male sperm and then the fertilized ovum splits into two. Identical twins look exactly alike. They have identical genes and are of the same sex.

Fraternal twins are formed when the mother releases two eggs (usually she releases only one egg per month) which are fertilized by two different sperm. The babies may be of different sexes, and they may look no more alike than any other brothers and sisters. But, of course, they share the same birthday.

Q *What are Siamese twins?*

A Siamese twins are two babies formed when the twinning zygote (fertilized ovum) doesn't separate completely. The two babies may be joined in some way, or they may share one organ between the two of them. Siamese twins are extremely rare. The most famous were Chang and Eng, born in Siam (now called Thailand) in the nineteenth century.

Q *Is it possible for a human mother to have more than two babies in her womb at the same time?*

A Yes, but it happens very rarely. The Dionne quintuplets (five children) were born in 1934. They came from a single egg that split into five identical parts. The quintuplets became famous all over the world.

Q *What are tears?*

A Tears are the fluids that bathe our eyes constantly. They flush out impurities. Every time you blink, tears clean out your eyeball. As far as we know, human beings are the only animals that shed tears, or "cry," when they become emotional.

Q *What are nerves?*

A The nerves that run throughout the body are like a telephone system that sends and receives messages to and from the brain.

It is through our nerves that the five senses function: we see, we smell, we taste, we hear, we feel.

It is because of our nerves that we feel pain. Pain is a danger signal. When you touch something hot or prickly, the nerves instantly send a danger-message to the brain, which tells your hand or foot to jump away from the source of pain. This nervous reaction happens automatically: you don't have to think about it.

NERVE CELL

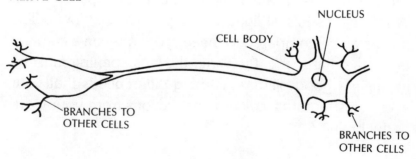

NUCLEUS

CELL BODY

BRANCHES TO
OTHER CELLS

BRANCHES TO
OTHER CELLS

Q *Why does pressing the area between your nose and upper lip prevent you from sneezing?*

A Because pressure at that point helps to deaden the nerve that receives the "sneeze" message from the brain.

However, it doesn't always work. If you have a heavy cold or an allergy, the brain insists that sneezing is necessary to get rid of irritating particles in the nose or throat.

Q *Why does the pupil of the eye sometimes seem bigger and sometimes smaller?*

A Because the pupil, which appears like a black spot in the center of the eyeball, is actually a hole that lets in light. When there isn't much light, the pupil opens wider. When there's lots of light, the pupil narrows down. The pupil may be compared to the lens opening on a camera: the darker the scene, the wider the lens opening must be to make a good picture. The big difference is that your eyes adjust *automatically* on instructions from your wonderful brain!

Q *What does 20/20 vision mean?*

A The expression 20/20 has come to mean "normal" vision. With 20/20 vision, one sees at 20 feet what the normal eye would see at that distance. A reading of 20/50 means that a person sees at 20 feet what the normal eye would see at 50 feet. The larger the second number, the poorer the eyesight. "Legal blindness" is defined as having 20/200 (or worse) vision in both eyes.

Q *What causes dizziness?*

A Some kinds of dizziness—the kind you get when you spin around and around and stop suddenly—is caused by the semicircular canals. After you've stopped spinning, the fluid in the canals continues to move for a short time. The nerve cells send confusing messages to your brain, and you feel dizzy. The dizziness ends when the fluid in the semicircular canals stops moving.

Q *How do your ears help you to walk?*

A By helping you to keep your balance. In the inner ear are organs called the semicircular canals. These canals form approximate right angles with each other. Two lie vertically and one is horizontal. The canals are lined with sensitive hairs and filled with fluid.

When the head and body move, the fluid also moves, but not quite as quickly as the rest of the body. As the movement bends the hairs in the canals, nerves send messages to the brain, which then corrects the balance of the body.

THE EAR

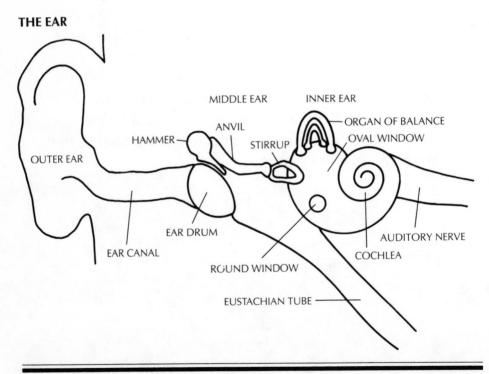

Without your semicircular canals you would have difficulty in knowing whether you were standing up straight, leaning to one side, or lying on your back or on your side.

Q *Does a deaf person have difficulty in keeping his or her balance?*

A No. The semicircular canals, which help us keep our balance, are not affected by deafness. You can test this by covering your ears and walking around.

Q *What causes snoring?*

A Snoring is the sound caused when air from the lungs vibrates on the uvula. If you open your mouth wide and look in the mirror, you can see your uvula: it is a tiny piece of flesh that hangs down from the palate (roof of the mouth) at the entrance to the throat. Snoring usually occurs when a person's nasal passages are congested and the person is breathing through the mouth.

Q *What causes dental cavities?*

A Cavities, or dental caries, are caused by microscopic bacteria that live in the mouth. The bacteria use foods such as sugar to make acid and a spongy substance called plaque. The bacteria live inside the plaque that coats the teeth, meanwhile attacking the tooth enamel with acid. Once a breakthrough has been made to the soft, inner pulp of the tooth, a cavity, or hole, is rapidly formed.

Brushing the teeth as often as possible helps to prevent the formation of plaque and to destroy already existing plaque.

Scientists hope that in the near future they will be able to produce a vaccine which will prevent the growth of *Streptococcus mutans,* the villainous bacteria that make their homes in our mouths.

SCIENCE AND TECHNOLOGY

From everyday objects such as electric light bulbs, telephones, and radios, to world-changing concepts such as nuclear power, robots, and computers, the world of science is all around us.

Experiments that once would have taken years to perfect and analyze are so speeded up by modern methods that it seems new discoveries are being made every day. More than ever before, our future success lies in being able to understand what is going on in the world of science.

Do you know about laser beams and radar? Do you know how the sound of your voice gets bounced off a satellite in space and back down to earth in seconds?

Find out the answers—and then keep checking for new discoveries!

Q *What is science?*

A The word "science" comes from the Latin word meaning "to know." People have always been curious about the world around them. Children begin asking questions almost as soon as they can talk. Adults may make up stories to explain things they don't understand. That is why there are myths and legends about the creation of the earth and other mysteries. But scientists look for answers, develop theories, and check them out, by making controlled experiments. A scientist is like a detective. He or she looks carefully at the subject and makes lots of notes. The object is measured, weighed, cut apart (if possible), described very exactly, compared to other objects that it may resemble, photographed, and perhaps tested in a hundred different ways.

Thanks to the invention of calculators and computers, science has made great leaps in the twentieth century, providing in moments answers to calculations or questions that would otherwise take weeks to obtain.

New discoveries are made every day. And new mysteries are created for scientists to investigate.

Q *What is an atom?*

A The atom is one of the tiny particles of which the earth—and everything in it and on it—is made up. And that includes you, your friends, tables, plants—everything!

An atom is so tiny that it cannot be seen without a very powerful microscope.

And yet an atom, we now know, is made up of even tinier particles. Each atom has a center (called a nucleus). The nucleus is in the center of a number of fast-whirling electrons.

For each electron, the nucleus contains a balancing number of protons. It is this balance of electrons and protons which gives the atom its particular character: what a collection of atoms looks like, how they behave near other atoms, and so on. For example, the gas oxygen has eight protons in it. An atom of the heavy precious metal, gold, has seventy-nine protons in its

THE ATOM

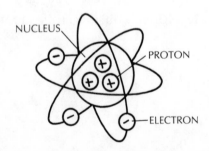

nucleus. The holding power between protons and electrons in an atom is extremely strong, despite the tininess of the particles.

Besides its halo of electrons and its body of protons, the atom also contains a neutron and other particles in its nucleus. Just how many tiny particles an atom contains is not clear at this time. New particles, and theories about how they go together, are being discovered all the time. It is possible that even a simple atom's nucleus may contain as many as forty-eight different particles. Keep checking!

Atoms bind together with one another in various combinations to form molecules of materials. An enormous collection of molecules forming the many ingredients of ink and paper is in the book that you are looking at right now.

Q *What is nuclear power?*

A Nuclear power is energy that is released when the central core of an atom (the nucleus) is split apart. The particles charge around, crashing into other atoms, which, if they are hit hard enough, in turn split apart, causing what is known as a "chain reaction." Such reactions can be very fast and extremely violent, and are the basis for atomic and hydrogen bombs.

A huge amount of power is released when atoms split, but it is not at all easy to maintain and control the release of this power. Gradually, scientists are learning how to use this huge amount of energy. It can be used to generate steam and thereby power generators that once used fossil fuels (such as oil and coal) as power sources. Nuclear power can have many uses, once we have learned how to control it.

Q *What is a laser beam?*

A A laser beam is an intense beam of light whose waves are all of a single frequency and all "in step" with one another.

The value of a laser lies in its light being in one small, thin beam. Instead of light being spread around in all directions, like that of a lamp, the laser beam is thin. All its energy is concentrated and controlled in a very tiny area.

Because of this kind of control, laser beams no wider than a hair can be used to perform delicate surgery on an eye. Or a laser can be designed to cut through steel sheets with intense heat. Laser beams have been aimed onto the moon and reflected back to earth in experiments to measure the speed and distance of the moon's orbit.

The ability to produce "cohesive" beams of light has helped engineers develop ways of sending signals (code, voice, television, etc.) in much the way that high-frequency radio waves travel.

Q *How may laser beams help communication between submarines and shore?*

A They can help because laser beams can penetrate water with immense energy at a tiny point and travel through it for miles without becoming distorted. Most conventional ways of sending sound, such as radio, have their signals dissipated and garbled after only a short voyage through water. In order to send or receive messages in the past, a submarine had to surface, thus revealing its presence to the enemy.

Now signals can be sent up to satellites which will then shoot the signal in pulses of blue-green laser light down to the submarine deep under the ocean.

Q *What does the word "laser" mean?*

A The word is made of the initials of the following words: Light Amplification by Stimulated Emission of Radiation.

A laser is a device in which certain atoms, when they are stimulated, or excited, by light waves or other energy sources, send out a very intense and focused beam of light.

The important thing about lasers is that the light emitted from it is "cohesive, which means "together."

Some materials give off light when they are subjected to intense, rapid sources of energy. The light comes from the changes in energy levels of the electrons as they jump into higher, and fall back into lower, orbits around the atom which they are circling. Usually the electrons are jumping and falling at random intervals, and light is given off all around, generally distributed.

But the laser is designed to subject its core material to a rapid, short, intense burst of energy. This causes the electrons to jump into orbit all together, that is, "cohesively." In this manner the laser material emits its light in a burst in one controllable direction.

Q *How does a radio work?*

A The radio with which we are most familiar is the receiving half of a radio system. The other half is the transmitter which sends out the signals.

At the radio transmitting station the messages which are to be sent out—music, voices, and so on—are turned into electrical signals which are impressed upon radio waves and sent out from the transmitter's antenna.

Radio signals are oscillating (vibrating) electromagnetic wave forms, and act just like light waves—except that we cannot see radio waves. Radio signals can travel through space at the speed of light, about 186,000 miles (nearly 300 million meters) per second, needing no wires or anything to conduct them.

Radio signals are sent out at many different frequencies (rates of oscillation), depending on the radio station that is transmitting them. At the receiving end, you can select your frequency, or favorite radio station, by turning the dial on your receiver.

Q *What do the numbers on a radio dial mean?*

A The numbers show the rate, or frequency, at which the electromagnetic radio waves are oscillating.

For example, the number "900 kHz" means that the signal is oscillating at 900,000 times per second.

In radio language, "k" stands for "thousand." "Hz" stands for "times per second," "m" stands for "million."

Another example: "108 mHz" stands for "108,000,000 vibrations per second."

The letters "Hz" come from the name of a German scientist, Heinrich Rudolph Hertz, who created much of the basic theory of radio.

Q *How does your radio receiver translate electromagnetic radio waves so that you can understand them?*

A Electromagnetic waves can be thought of as packets of energy moving through space, carrying the messages impressed upon them. The job of the radio receiver is to catch those packets of energy and turn them back into messages that you and I can understand.

The electromagnetic radio waves, striking the antenna of the receiver, create minute electrical signals.

The receiver amplifies those signals, selecting the ones to which it is tuned and rejecting the others. It then turns the radio signals back into understandable, audible signals that can be heard through the radio's loudspeaker.

Q *How does a telephone work?*

A Today's telephone system is an intricate and complex one that stretches around the world. But its basic elements are simple, so let's start there.

Imagine that we have two telephones connected directly to one another by wires.

The telephone part that you talk into is called the transmitter. When you speak into it, a thin metal plate called a diaphragm is ready to receive your voice's sound waves. The sound waves cause the thin metal to vibrate. The vibrations pass through into a cup containing carbon, which is carrying a small electrical current. Depending on the "frequency" of your voice, the carbon molecules are made active at greater or lesser speed, affecting the electric waves that then pass through wires connecting with a receiver at the other end.

The electric current activated by your voice goes through another thin metal diaphragm, making it vibrate, creating sound waves. The listening ear receives sound waves that the brain translates into words—and all of this happens in thousandths of a second!

Q *What is the telephone "switching system"?*

A It is a complex system by which telephone messages can travel for thousands of miles. There are at least 100 million telephones in the United States, each of which can be connected to any other.

Each town has a central switching system station run by the telephone company. Your phone, which has both a transmitter and a receiver in it, connects to the central station, as do all the other phones in your neighborhood.

When you pick up the phone and dial a number, your dial sends a series of pulses (or tones on a push-button phone) to the central station. The central station receives the pulses, and, in a modern switching center, a computer-driven system will understand the code of pulses and make connection with the telephone number you have dialed. If the call was out of town, then your telephone signals would be routed in this manner: to the local central switching center; to the trunk link to the out-of-town central switch; to the out-of-town telephone.

Q *How can telephone messages be sent all the way around the world in seconds?*

A Today's telephone technology is so advanced that the trunk link between central switches may be a microwave radio link; or a laser light beam carrying thousands of calls down a tiny strand of glass; or a satellite link beaming your telephone voice by radio to a satellite thousands of miles above the earth, retransmitting it down to earth, and receiving it on the other side of the continent or across the ocean. The days of telephone wires strung from telephone poles have now all but disappeared.

Next time you call overseas or long distance, try to tell whether your voice is traveling by way of a satellite. If it is, you can sometimes hear a faint echo of your voice delayed a second or so by the travel time of the radio waves, just recognizable in the faint noise you can hear in your phone's earpiece receiver.

Q *What is the difference between the telephone-wire system and radio?*

A The difference between the telephone-wire system and radio is that a radio transmission needs no wire to conduct it to the receiver. When radio was invented, it was called "the wireless," its lack of wires being the special feature that distinguished it from signals sent by wire. The radio system uses electrical energy which is different in form from the electrical current carried along a wire. Radio signals are electromagnetic (from the words *electricity* and *magnetism*). Electromagnetic systems radiate in space at the speed of light. In fact, they behave very much like light, able to travel between galaxies and stars.

Q *How can remote places, such as jungle areas and mountaintops, receive telephone service?*

A With the help of a new device called a solar-powered repeater. The repeater gets its power from the sun, so no telephone lines or cables are necessary. Instead, telephone calls are relayed by radio wave from one repeater to another, until they eventually reach their destination.

   To install repeater stations, teams of people and equipment must be flown to the inaccessible places, often by helicopter.

Q *How does a stethoscope work?*

A The stethoscope works by concentrating and magnifying sound. It forms a direct line between the body and the listener, who is usually a doctor. One end of the stethoscope is a funnel-shaped rubber cup. The cup clings close to the body, shutting off outside sounds. There are two tubes coming from the cup. The doctor puts one tube into each of his ears. Now the doctor cannot hear any sound except those within your body. The doctor can hear the heart beating, the lungs breathing, and the digestive system digesting.

By the way, the word "stethoscope" comes from two words, *steth,* Greek for chest, and *scope,* Latin for watcher. A stethoscope is a chest watcher.

Q *What makes an electric light bulb glow?*

A The light bulb glows when a switch is pressed, allowing an electric current to pass through a filament (a very fine metal wire) that is a good conductor of electricity. The electric current causes the molecules of the filament to move so rapidly that they get white hot and give off light.

Q *Who invented the first electric light bulb?*

A The first workable incandescent (glowing) bulb was developed by two people on different sides of the Atlantic Ocean at about the same time: Sir Joseph Swan in England and Thomas Alva Edison, in 1879, in the United States.

Today, light bulbs come in all shapes and sizes. Some—for example, those used in medical instruments—are no bigger than a grain of rice. Others, such as those in a lighthouse, are as big as a football.

Q *What does the word "scuba" mean?*

A The word is made up of the first letters of the words Self-Contained-Underwater-Breathing-Apparatus. Scuba gear is also called an aqualung (water lung).

Q *How does scuba gear work?*

A Scuba gear works by allowing a diver to carry his or her own supply of air underwater. Scuba consists of one or more strong metal tanks containing air. These tanks are strapped to the diver's back. The diver inhales air from the tank through a tube attached to the tank. A regulator valve carefully balances the water pressure against the air pressure in the diver's tank, so he or she is able to breathe normally.

The exhaled air leaves through another tube in the form of bubbles. A pressure gauge shows the diver how much air is left in the tanks.

Q *Why and how does water pressure affect a person's body?*

A Water is heavier than air. Human beings are used to the weight, or pressure, of air. But when they go underwater, the heavier pressure of water can be felt in air cavities such as the lungs, sinuses, the inner ear, and the stomach. These air spaces can get squashed in, just as the pressure of your hand on an air-filled balloon can gradually squeeze in and crush the balloon, making it explode.

The "demand valve" on scuba gear regulates the amount of air breathed in and out (through a tube in the mouth) and equalizes the pressure between the cavities of the body and the weight of the surrounding water, and so the diver can breathe normally in spite of the great pressure on his or her body.

Q *How do the bubbles from an aqualung (or scuba) help a diver to know which way is up?*

A Because air bubbles are lighter than water and always rise upward, toward the surface. By looking at the direction in which the exhaled bubbles are traveling, the diver always knows which way is "up." This knowledge is important to a diver, who may easily get confused about directions in the weightless underwater world that is so strange to human beings.

Q *What is the northernmost shipwreck to be explored by man?*

A The shipwreck is H.M.S. *Breadalbane* (pronounced bread-ALL-bin), a 125-foot sailing vessel that sank into the subfreezing waters of the Canadian Arctic in 1853. Today the vessel is remarkably well preserved. In the deep-freeze of Lancaster Sound, 600 miles north of the Arctic Circle and only 100 miles from the North Pole, there are no sea worms to eat at the wood, and no swift-moving currents to help crumble the vessel.

Q *Why is the exploration of the* **Breadalbane's wreck** *so important to scientists?*

A Because this frozen "time-capsule" will tell scientists much about biology (living creatures), geology (the structure of the sea bottom), and sea ice. All of this knowledge will be most valuable as the Arctic is being probed for oil and gas for the future.

The divers exploring the wreck have already sent back remarkable television pictures of the wreck and its surroundings. The images are sent by satellite to science centers, where they can be studied in comfort, far away from the ice pack.

Q *What "superbug" has been "invented" that can help clean up oil spills?*

A A strain of the bacterium *Pseudomonas.* This microscopic creature was created in the laboratory by combining the genes of several bacteria that devour different chemicals in oil.

The invention marked an historic occasion. In June, 1980, the U.S. Supreme Court ruled that genetically engineered bacteria could be patented.

The name of the scientist who won the patent: Ananda Chakrabarty, University of Illinois.

Q *Why do you get an electric shock if you walk across a rug and then touch a metal doorknob or another person?*

A Because the scuffing of your feet along the pile of a carpet sets free some electrons from the carpet and you get charged with static electricity. When you touch an object (particularly a good conductor of electricity, such as a metal doorknob), the electrons leap from you and cause a tiny spark.

The same thing happens when you rub a balloon against a furry sweater: the balloon gets some electrons from the hair. When you set it against the wall, the charge of electricity from the balloon will make it stick to the wall for a short time.

Q *Why did scientists send live frogs on a space shuttle flight?*

A Because they wanted to study the effect of zero gravity on developing frog embryos. In its normal pond habitat, a fertilized frog egg undergoes important changes caused by the earth's gravity, which makes the egg rotate and develop. Without the normal rotation, the rapidly dividing cells may not go where they should, to form two eyes, four legs, and so on. The results of the experiment may help scientists to form new ideas about the biology of other animals, including humans. From studies such as these, environmental systems capable of maintaining people safely in space are developed.

Q *How does a computer "think"?*

A The computer "thinks" by recording and operating upon its inputs and recorded programs in a series of electric pulses called bits. At its simplest level, when a pulse is present, the computer registers 1 (one bit). When there is no pulse, the computer "thinks" 0 (zero bit). The way in which the computer handles its bits of information is called the program.

Computers can think their way through thousands of bits in a second, without getting tired and making mistakes. This makes computers very useful for doing things that humans are slow at, or don't want to do—things such as adding and subtracting.

Q *Could a computer, or any robot, become more intelligent than the people who program it?*

A Maybe. It makes good science fiction to imagine that a machine can beat humans at thinking. Certainly a computer can tell you when you have made a mistake. But a computer is still only a machine—though a most remarkable one. It can think faster than a human mind under certain conditions, and it can have a more reliable and larger memory of facts.

For example, a computer could have all the information contained in all the books of a huge library in its memory. It could be programmed to search out certain sentences from its vast memory in a few days. By contrast, it might take a human being a lifetime of reading to find just one sentence—unless that person were able to bring other ideas to help in the search: childhood memories, stories and plays, places and people. The random searching of a human mind, with all its associations, can sometimes beat a computerized "brain."

But computer science is at a stage when computers are being programmed to look for the same clues that a human might use. If and when such programming is achieved, we'd better watch out!

Q *What is the difference between film and videotape?*

A Film is a transparent plastic material coated with light-sensitive chemicals designed to capture a visual image.

To see a film when it has been exposed, it must be developed with the use of chemicals. Films form a positive image, such as a slide, that can be seen with the naked eye or through a projector, or it can be made into a negative from which positive prints can be made.

Videotape works by electronic means. Videotape is a plastic material coated with a magnetic surface. Videotape can be viewed only with electronic equipment capable of turning the information which is magnetically stored on the tape into a signal usually viewed on a TV receiver.

Film and videotape cannot be projected through the same equipment. Videotape can be erased and used again, while film cannot.

Q *What are X-rays?*

A X-rays are types of energy, rather like light rays. X-rays can travel through empty space at close to the speed of light. They can also travel through materials such as wood or flesh, but are usually stopped by very dense and

thick things such as lead and steel. This ability to travel through materials that light cannot penetrate makes them most useful in medicine. You have probably seen a "shadow picture" of your teeth or lungs or a leg or arm bone in your doctor's office.

Q *How do X-rays work?*

A X-rays will travel through your body and leave a trace on a sensitive X-ray plate. The dense parts of your body absorb the X-rays' energy, and the less dense, fleshy parts let more X-rays pass through onto the plate, so the radiographer can develop a "shadow picture" of your insides.

Q *How are X-rays different from light rays?*

A We cannot see X-rays. Our eyes are not sensitive to X-rays, as they are to light rays.

Q *What is that CAT doing in a hospital room?*

A It is helping doctors diagnose, or find out, a patient's illness. The letters CAT stand for *Computerized Axial Tomography.* The CAT scanner rapidly rotates a single X-ray beam around a human head or body, and the varying thicknesses and densities of flesh, organs, and bones absorb the X-rays at different rates. The X-rays that pass through the body hit X-ray detectors which produce electrical signals of varying strengths, depending upon the amount of the X-ray that gets through. In turn, the signals are fed to a computer which is programmed to undertake the very difficult task of interpreting the information it receives.

The outcome is that the computer is able to reconstruct a very accurate picture of your insides, and show it to the doctor on a TV screen.

The CAT scanner represents a great advance in medical science, since it doesn't subject the patient to multiple and dangerous X-rays.

Q *What is radar?*

A The word *radar* is short for *Radio Detection and Ranging.*

Radar equipment makes use of the facts that radio waves can travel through space in straight lines and can be deflected by dense objects, such as metal airplanes, steel ships, missiles, tall buildings, towers, and hills.

The radar equipment is designed to send out a pulse of radio waves in a given direction; then, for a fraction of a second, a receiver listens to hear if any signals are reflected back from where the pulse was aimed. This sequence of events is repeated over and over again, with the radar-scanner system looking at all points of the compass in turn. It takes only five or ten seconds to search completely around the horizon.

If the receiver senses any of the transmitted radio signals coming back to it, a computer calculates quickly how long it took for the radio pulse to go out and be received. Because it knows the speed at which radio waves travel, it can determine how far away the object is which reflected the signal.

Now the radar system knows two things. It knows in what direction it sent out the pulse, and how far away the object is. The radar system shows the operator this information on a radar screen, by writing a small blip of light on a circular plan of the area around the radar scanner.

Let's say an airplane is flying through thick fog. The pilot cannot see anything ahead of him. But if he turns on his radar screen, he will see little points of light on the screen. Each point of light, or "blip," represents a signal pulse of an object that has been bounced back from his radar transmitter. Now the pilot knows whether or not he must change direction to avoid a collision.

Although radar systems are usually designed to look for metal objects, they can be made to sense rain and storm clouds. Sometimes a radar system will even spot the moon peeping up over the edge of the horizon! But most systems have a range of only a few to a hundred or so miles.

Q *Why does soda bubble after its container is opened?*

A The bubbling is caused by the gas, carbon dioxide, that is suddenly released from pressure. Soda water, like other "carbonated" liquids, contains carbon dioxide that has been dissolved under pressure. When the pressure is released by opening the container, the liquid cannot hold as much carbon dioxide, so the excess bubbles out.

Q *Why does a bottle or can of carbonated beverage get "flat" after it is opened?*

A It goes "flat" because the carbon dioxide that formed the bubbles escapes into the air. If you want your drink to lose its bubbles faster, stir it so the bubbles of carbon dioxide can escape faster.

Q *What is glass made of?*

A Glass is commonly made of three ingredients: sand, soda ash, and lime, melted together at high temperatures. The melting is done in large furnaces. The mixture melts into a syrupy mass which, when cooled, becomes glass.

Q *How is glass shaped into many different objects?*

A Glass is shaped when it is still in the "syrupy" stage. One way to form it into, let's say, the shape of a bubble is called "blowing." The syrupy mass is poured into an open-ended tube. Then either a person or a machine blows into the tube, and a shape is formed. (You can get an idea of how this happens by blowing a soap bubble.)

These glass bubbles are used as containers such as bottles, jars, drinking glasses, pitchers, and so on.

The soft glass mixture must constantly be reheated to keep it soft and workable until it is the right shape.

Then it is gradually cooled, reheated, and cooled again. This process makes the glass hard, and is called annealing.

Q *Is it possible for an animal to give birth to another animal of a different species?*

A Yes. In August, 1981, Flossie, a Holstein cow, gave birth to a healthy gaur calf. A gaur is a wild ox, native to India, and an endangered species. Scientists took a fertilized egg from a gaur's body and injected it into the uterus of the cow at the Bronx Zoo in New York. The new baby weighed a healthy 73 pounds at birth. Its proud keepers gave it the name Manhar, which in Indian means, "He who wins everyone's heart."

The birth of Manhar was the second successful transplant of an egg to take place in the United States. The first event of this kind was at Utah State University in 1977, when a baby mouflon (a wild Sardinian sheep) was born to a domestic sheep.

Successful embryo transplants mean that rare and endangered species will have a better chance of survival.

THE
SUPERNATURAL

Long ago, when our ancestors lived in isolated communities and the nights were long and dark, the world was full of mysteries. People truly believed in witchcraft and ghosts, vampires and werewolves.

Today we may laugh at ancient superstitions—but we have our own mysteries, such as those of extra-sensory perception, unidentified flying objects, visitors from outer space, the Bermuda Triangle.

Do stories of the supernatural have scientific explanations?

We may never find out—but most of us are still fascinated by spooky stories. Here are some questions and answers to chill your spine.

Q What is a vampire?

A In folklore, a vampire is a dead person who returns from the grave. Vampires are said to come out of their burial places at nightfall and to feed on human blood that they need to survive, sucking the blood from their victim's neck. According to superstition, a person bitten by a vampire also becomes a vampire. Sometimes the vampire may assume a nonhuman shape, such as that of a bat. Other beliefs are that the vampire has no reflection in a mirror, and that it must return to its coffin before daylight. It can be killed only by having a wooden stake driven through its heart.

Q Do vampires exist?

A As far as we know, they exist only in people's minds.

But the belief about the dead draining blood from the living is deep-rooted in folklore all over the world, and is persistent, even today.

Some people believe in vampires so strongly that they sleep with cloves of garlic around their necks, and with salt sprinkled on the windowsill. Both of these substances are supposed to be repellent to vampires.

Some people go to their church to have their fear of vampires exorcised; some may go to psychiatrists. There have even been "vampire hunters" who have dug up corpses (at Highgate Cemetery, London, 1974) and beheaded them or driven stakes through their hearts.

Q *Why do people believe in vampires?*

A One possible reason for the origin of the belief is that in ancient times, when people didn't know much about the workings of the human body, it was quite common for patients in a coma or even in a deep sleep to be put into coffins because they appeared to be dead. If they revived in time to escape, empty coffins and opened burial vaults were found, and people assumed that the "dead" had arisen. There are many tales told about corpses being found with bloodstained mouths and hands when their coffins were opened, for one reason or another—often by grave robbers. The blood was doubtless the result of the victim trying to claw his way out before dying of suffocation.

During the Middle Ages, when the dreaded plague, or "Black Death," swept Europe, killing thousands, terrified people left their dead and dying and fled. Live bodies were sometimes piled onto carts along with the dead and brought to the graveyard or common burial pit. When some of the "dead" screamed and leaped off the death wagon or moved out of their intended burial places, it is no wonder that superstitious people began to believe in "the living dead."

In modern times, plays, movies, and books about Count Dracula and others have been responsible for the perpetuation of the vampire legend.

Q *Who was Count Dracula?*

A Count Dracula was a famous vampire conceived by author Bram Stoker, in 1897. The story described Dracula as tall and thin. He dressed in black and wore a cape. When he thirsted for blood he sprouted long, sharp teeth. He could change into a bat or a wolf. Dracula lived in a castle in Transylvania, which is now part of Romania in eastern Europe.

Bram Stoker based his story on that of a real person called Vlad the Impaler, a fifteenth-century ruler in eastern Europe. He was the son of Prince Vlad Dracul (Vlad the Devil) and was therefore also known as Dracula (son of the Devil). Vlad was cruel to his enemies, but he was a good ruler and today is a national hero in Romania.

Q *What is a werewolf?*

A The word comes from the Old English, "wer," meaning man: "man-wolf." Legends tell us that if a person, usually a man, but sometimes a woman, enters into a pact with the Wolf Spirit or with the Devil, that person will turn into a wolf at sundown and resume human shape during the day.

  The werewolf hunts at night, often attacking and eating people, particularly children. If it is wounded, it immediately becomes human again. It is best killed with a silver bullet or by being burned at a stake.

Q *Why would anyone want to become a werewolf?*

A Because a wolf, and other large, powerful, hairy meat-eaters were figures of terror in ancient times. People thought that if they could become fierce animals they would become strong, powerful, and terrifying. They could do evil things and not get caught.

  However, there are many, many records of werewolves being caught and brought to trial. There was an epidemic of werewolves in France in the sixteenth century. Sometimes the court decided that the person was a victim of a mental disease called lycanthropy, in which the person suffers from the delusion that he or she is a wolf. In that case, the person might be released as merely insane.

But more often than not, the killers were horribly punished.

Q *How did stories about men turning into animals begin?*

A We don't know for sure, but it seems likely that they began a very long time ago, when hunters would disguise themselves in the skins of animals to stalk their prey. Only a few hundred years ago white settlers saw and drew pictures of native Americans, clad in buffalo skins, hunting deer and buffalo. Since they crouched on all fours and wore fur skins and moved slowly, they were able to creep up on their victims more easily than as light-skinned, upright, two-legged men.

Barbaric warriors, too, covered themselves in the furry skins of animals. Often they would work themselves into a frenzy of screaming, prancing, and growling while attacking terrified peasants, who mistook them for animals.

To this day, there are secret societies of leopard-men in Africa, jaguar-men in South America, crocodile-men, bear-men—all of these imitate large, powerful, intimidating creatures and strike terror into the hearts of most people.

Q *What are ghosts?*

A Ghosts are said to be the restless spirits of the dead. They may be seen as transparent, formless wisps, or as solid, recognizable forms. Sometimes there are accompanying sounds, from eerie moans to voices to the galloping of invisible hooves.

People who believe in ghosts say that they are most often the spirits of those who have died sudden or violent deaths. Sometimes an apparition is felt to be an omen of bad luck. Sometimes the ghosts of saintly people are taken to be divine messengers from above.

Often the ghosts are associated with local history and superstitions. There are "haunted" houses, "family ghosts," "haunted" forests and mountains, and countless ghost stories from all kinds of people in all parts of the world.

Q *What is a poltergeist?*

A It is a particularly noisy, and often mischievous, kind of ghost, usually heard and felt but not seen. The word comes form the German, "knocking spirit," which is a good description. The poltergeist makes rapping noises, moves furniture, sends things flying through the air, breaks glass and china, and generally misbehaves.

Unlike more delicate spirits, poltergeists leave definite evidence of their presence—broken crockery and up-turned chairs—for all to see. But though they have been "investigated" by modern parapsychologists (people who study supernatural happenings), there is no real proof that they exist. And no reasonable explanation can be given for their antics.

Q *Do ghosts really exist?*

A Most people would say, "No, of course not. They are figments of an overactive imagination, or illusions caused by the physical or emotional state of the viewer, or by clever fakeries manufactured for the benefit of those who *want* to believe." But these skeptics have been shouted down by "believers" for centuries, and both will probably continue to argue, for the truth is we may never know the answer to the question.

Q *Why do people dress up as ghosts and skeletons on Halloween?*

A The word *Halloween* comes from "Hallows Eve," that is, the evening before All Saints' Day (November 1). Way back in the seventh century it was decided by the Roman

Catholic Church that on this day all saints, martyrs, and other holy (or hallowed) people should be remembered in religious ceremonies.

But the customs of Halloween (the scary ghost stories, the jack-o'-lantern, the dressing-up) are Celtic and pagan in origin, dating back to earliest history. Long before calendars were invented, people in cold northern climates decided that the time of the year around November 1 was the beginning of winter—a time of long, dark evenings (with no light but firebrands), bitter cold, and little food: in fact, a perfect time for evil spirits to wander around. Accordingly, primitive people dressed themselves up to frighten away evil spirits—and to give themselves courage.

In many parts of the British Isles the custom of lighting huge bonfires and gathering around them still continues. In the United States, children dress up and go out "trick-or-treating."

Q *Who are fairies, goblins, and other "little people"?*

A They are, as far as we know, imaginary creatures invented by storytellers. They have been present in folktales all over the world for thousands of years. Some stories even say that at one time there may have been a race of small people who took to the deep forests, mountains, and caves when bigger people evolved on earth. Could it be that the descendants of tiny folk are still around carefully hidden? After all, most of the upright hominids were under four feet tall, and present-day pygmies in Africa are not much bigger.

Q *What is a will-o'-the-wisp?*

A It's a dancing light that appears over marshes and other damp ground. For a long time people thought that it was an evil spirit, trying to attract people to be drowned in dangerous places. Now scientists think it is probably the natural gas, methane, that has been lit by electric currents in air and water vapor. The light caused by methane quickly disappears, but another bubble of gas may reappear somewhere else. Anyone watching could easily think that the dancing lights of methane were caused by supernatural creatures.

Q *What is voodoo?*

A Voodoo is the folk religion of the island of Haiti in the Caribbean. It originated in Africa, but is now mixed with local Carib Indian customs and Christian church elements.

There is much chanting, dancing, and beating of drums at a voodoo ceremony. Sometimes people go into a trance and believe that they are "possessed" by a spirit, called a loa. The loa can command the possessed person to perform strange or evil acts, such as stealing, lying, or even killing.

Q *What is a zombie?*

A In the voodoo religion, a zombie is a "living-dead"—a person recalled from the dead by a sorcerer in order to perform certain tasks, especially wicked deeds such as killing.

Q *What does it mean when something is "taboo"?*

A It means, "Don't do it," or "Don't touch it." It comes from a Polynesian word meaning "set apart." In some Polynesian societies and, in fact, in nearly all cultures there are beliefs or superstitions dating back to ancient times when people believed in magic. Doing something or touching something taboo could mean that something bad might happen to you or to your tribe.

Q *Can death be caused by an evil spell?*

A In some religions, such as voodoo, people believe that it can. They believe that sticking pins into a doll or, these days, into a photograph of a person, can cause death. There is no direct evidence that death can be caused by sorcery. Modern physicians think that a person who believes himself to be under a spell may actually cause his own death by refusing to eat or drink, and in fact may die of fright, or of some already existing disease, or of poison administered by the sorcerer.

Q *What is demonic possession?*

A In religious belief, it is a condition in which a person's mind, body, and soul are taken over by an evil spirit or demon.

The possessed person is frightened of religious symbols, such as crucifixes. He or she acts in a strange manner: a totally unknown language may be spoken in a voice not resembling that of the "original" person. The victim may go into convulsions, start screaming, laughing, or throwing things around, and generally be destructive.

In modern times, a psychiatrist (a physician who specializes in the workings of the human mind) may be called in to find out if the possessed person has schizophrenia or some other disorder of the mind.

But even in modern society it is often the exorcist, a religious specialist, who is asked to exorcise, or cast out, the devil. If the victim and the supporting family, have faith in the practice, the efforts of the exorcist may be successful. Most scientists believe that demonic possession is the result of a severe psychological disturbance and can be cured by psychotherapy.

Q *Can people really predict the future by gazing into crystal balls, reading palms, and looking at the tea leaves left in a cup?*

A There is no scientific evidence to prove that "fortunetelling" is possible. But thousands of people have believed in it, and still do. In ancient times prophets and magicians sacrificed animals and studied their entrails (inner organs, such as heart, liver, and intestines).

A person gazing into a crystal ball may easily go into a form of self-hypnotism, brought on by an interplay of lights and reflections caused by the glass. People in such a state of trance may say things that, by chance, turn out to be true.

The fact that the marks on the palms of your hands are unique—unlike anybody else's—has long made people think that the lines mean something special. Palm readers, and in fact most clairvoyants (people who say they can predict the future) are skilled and intelligent people who can tell a great deal about a person simply by observing his or her appearance and listening to speech patterns. They may make very accurate statements about a person and his or her way of life and what is likely to happen to that person. The terms used are rather general (for example, "You are going on a journey,") and so could apply to almost any situation. Similarly, with tea leaves, much depends on the creative imagination and acute observation of the fortuneteller and the readiness of the customer to provide clues by reactions and responses, along with a desire to believe. Not many people truly believe in fortunetellers—they just go along with the act and then forget it.

Q *What is E.S.P.?*

A It is short for extrasensory perception. "Extrasensory" means beyond the five senses that we all have: the senses of sight, hearing, smell, taste, and touch.

Q *Does E.S.P. really exist?*

A Nobody knows for sure. For thousands of years people have believed that there is a sixth sense possessed by certain respected or feared members of the community: witches, medicine men, seers, and, more recently, psychics (who are ordinary people that seem to have extraordinary powers). In the past, the powers of these people were taken for granted.

In our present age of technology, belief in psychic phenomena seems ridiculous to many people. However, serious scientists such as Dr. J. B. Rhine (Duke University, North Carolina, 1947,), and Drs. Puthoff and Targ (at Stanford Research Institute, California, 1977), and others have tried to prove the existence of E.S.P.

These scientists reported that they performed thousands of carefully monitored experiments with scores of people under laboratory conditions. They were satisfied that E.S.P. does indeed exist and that anyone may experience it.

However, other scientists feel strongly that the experiments were unconventional and unscientific, and that much more research must be done on this highly controversial subject.

Whatever the scientists do or say, there are few people who haven't had an apparent extrasensory experience at one time or another. As with any happenings that are out of the ordinary, there are those who dismiss both the stories and the experiments as chance or coincidence; those who are fanatical believers; and those who keep an interested "wait-and-see—and keep-checking!" attitude.

Q *What is witchcraft?*

A It may surprise you to know that many people, even today, consider it as a form of religion—"the oldest religion," they call it.

There seem to be many kinds of witchcraft, some good, some evil.

"Good" or "white" witches are a modern offshoot of witchcraft. These witches (both men and women) hold their meetings in each other's houses, are interviewed by newspapers, and even appear on television. Some of them help parapsychologists (people who investigate supernatural happenings) in their work. White witches believe that they have special powers to do good and help people.

At its worst, witchcraft is called Satanism, or Devil Worship. Its members are sworn to secrecy and threatened with gruesome punishment if they try to leave the coven (group). They believe in the power to do evil. Psychologists say that these are mentally disturbed people and often criminals.

Q *Why do people believe in witchcraft?*

A Because people have always had a need to believe in some kind of magic or supernatural power.

Long before today's religions took root, primitive people believed in good and evil forces at work around them. In a world that could often be hostile, people needed an ally, or even better, a means of obtaining power and knowledge to do good or evil themselves.

People who were (or are) thought to have special powers are given various names: witches, sorcerers, magicians (mostly in Europe), and medicine men, witch doctors, and shamans, in Africa and the Americas, to name a few.

Q *Do witches still exist?*

A People who claim to have magic powers still exist, though they are not necessarily called witches. There are the "white witches," already mentioned. There are "faith healers" who, some believe, can cure the sick; and, especially in primitive societies, there are the "witch doctors" who can cast spells for good or evil.

In spite of all the efforts of modern science to disprove the existence of supernatural powers, many people believe in all kinds of magic, from the superstitions of our everyday world (such as "wishing on a star") to sincerely practiced religions (such as voodoo).

Q *Why were witches punished and burned at a stake in the Middle Ages (and even later)?*

A The Church had gained tremendous power in Europe in the Middle Ages. But many people still believed in ancient sorcery and often secretly sought the help of witches. The Church considered witches not only as evil people who worshiped the Devil, but as rivals to the power of the Church. They sought out and cruelly punished people thought to be witches.

Witches became scapegoats, that is, they could be blamed for anything bad that happened, from a death to a poor crop to a burning house. Unfortunately, people were quick to accuse their enemies of being witches, and many innocent people were put to death. (The last known witch hanging in this country was in Salem, Mass., in 1692.)

Q *How did the image of an old hag riding a broomstick with a cat come about?*

A This image of a witch was an invention of superstitious peasants and imaginative storytellers.

People were afraid of witches. They truly thought that witches could transport themselves from one place to another by magic—so why not on a broomstick, at that time a common household article.

Cats to this day are considered somewhat mysterious. (As anyone who has lived with a cat knows they have a talent for appearing in odd places and disappearing when we are not looking. They also get "spooked" and behave as if they can see or hear something that we cannot.) People came to believe that witches could turn themselves into cats. Since witches practiced "black magic," it seemed reasonable that they should dress in black and have black cats. Cats came to be called "familiars" of witches. (Other animals, such as goats, were also associated with witches, but the superstition of a black cat being a sign of good or bad luck persists to this day.)

Q *What is a well-known witch's recipe in literature?*

A The following recipe was chanted by three witches stirring their kettle (or cauldron) in William Shakespeare's famous play, *Macbeth*.

> Double, double, toil and trouble;
> Fire burn and cauldron bubble;
> Fillet of a fenny snake,
> In the cauldron boil and bake;
> Eye of newt and toe of frog,
> Wool of bat, and tongue of dog,
> Adder's fork, and blind-worm's sting,
> Lizard's leg, and howlet's wing,
> For a charm of powerful trouble,
> Like a hell-broth boil and bubble.

If you want to read more of this tasty recipe, look up *Macbeth*, Act 4, Scene 1.

Q *What, if anything, is a dragon?*

A A most fearsome creature that plays the monster role in hundreds of fairy tales and folk stories all over the world. It is usually pictured as a large, long, creature with a spiky tail, a scale-coated skin, a large, toothy mouth, and nostrils that breathe flames and smoke. In ancient stories the dragon is usually slain by a brave knight in armor.

Q *How did people come to believe in giants?*

A Ancient peoples didn't know as much as we do about themselves, the world around them, and the forces of nature. They imagined that for huge mountains and rivers to exist, and for thunder and lightning and volcanic eruptions to occur, some being much larger than they must be at work. Ancient Greeks and Romans believed that mighty gods and goddesses ruled the world from the great heights of Mount Olympus, a real mountain.

When ancient peoples found huge bones, they believed that they were the bones of giants who had lived long ago. Now we know that the bones were probably fossils of large animals such as wooly mammoths that lived centuries ago.

Q *Did dragons ever really exist?*

A Not as far as we know, but a possible explanation for their presence in so many stories is this: people have always been frightened by snakes, lizards, and crocodiles, even though many of these creatures are harmless. To give themselves courage, maybe they invented creatures that were scarier and more powerful than the real thing, and invented powerful heros that conquered the monsters.

Q *Which people still feature the dragon in a yearly festival?*

A The Chinese. At Chinese New Year (which lasts a whole month, from the first new moon of January until the next full moon), part of the fun is to help conquer the "monster." A line of people, covered with cloth or fire-proof paper, make up the fire-breathing dragon. Lots of noise from firecrackers and lights from lamps are supposed to frighten the monster away for a whole year—and besides, everyone has fun!

Q *Is the Komodo Dragon really a dragon?*

A No. It is a very large lizard (about 10 feet long and weighing over 300 pounds.) that still lives on the island of Komodo, Indonesia. This lizard, or one of its relatives, could have served as a very good model for people's ideas about dragonlike monsters. It is long with a massive tail, the usual reptilian, scaly skin, and heavy, clawed feet. It does not, however, breathe fire.

Q *Who or what is the Loch Ness monster?*

A "Nessie," the popular nickname for the Loch Ness monster, is a mysterious creature that has puzzled and fascinated people for many years.

Loch Ness is a large, deep freshwater lake in Scotland.

Ancient documents show that for hundreds of years people have reported seeing a huge, long-necked creature in the dark and often misty waters of the lake. In recent years people have even taken some blurry photographs of what is supposed to be the monster. It looks somewhat like a plesiosaur—but plesiosaurs, like dinosaurs, died out millions of years ago.

Some people think that the "monster" may be a string of otters swimming together and darting in and out of the water. From a distance, they could give the appearance of a single creature rippling through the water.

Nevertheless, looking for Nessie has become an annual hobby for thousands of people.

Q *What is the Abominable Snowman?*

A It is said to be a creature that lives in the remote and icy wastes of the Himalayan Mountains in Asia.

People who have seen it describe it as being very large, half man, half ape. It leaves enormous footprints in the snow—about 12 inches long and 7½ inches wide—bigger than those of the biggest bears or apes. It has five toes, and seems to be flat-footed.

Local people call it a Yeti. It has been in their folklore for hundreds of years and seems remarkably like "Big Foot," also known as "Sasquatch," who appears in the folklore of native North Americans.

Q *What is Big Foot?*

A Big Foot is a very large creature that walks on two feet, seems to be manlike and apelike, and roams the deep northwest forests of western North America, from British Columbia all the way down to Central America.

If it exists (and there is no scientific evidence that it does), it is remarkably like the Abominable Snowman, or Yeti, of the Asian Himalayan Mountains. It certainly appears in the folklore of both Asians and native Americans.

Since we believe that native Americans first arrived in America by crossing the Bering Strait from Asia, it is not surprising that both groups of peoples have the same folklore. But is it a legend, or is it fact? Nobody knows for sure.

Q *What are the mysterious "Nasca Lines" of the Andes?*

A They are a series of markings north and south of the town of Nasca, about 200 miles south of Lima, Peru. The mystery about them is that they can be seen only from a height of about 1,000 feet up in the air, unless you happen to be standing directly astride one of them.

The markings consist of perfectly straight lines where the blackened stones that lie on the ground have been scraped aside. There are also markings in rocks.

The lines appear to be about one thousand to three thousand years old, though nobody noticed them until airplanes started flying over this desolate area. They don't seem to be roads, as they don't lead anywhere.

Imaginative writers have speculated that the lines were once made by visitors from outer space, who used the lines to guide their spaceships onto the "landing strip."

Others think that the lines were made as ceremonial pathways for some ancient Inca ritual.

Perhaps the mystery will never be solved.

Keep checking!

Q *What is the Bermuda Triangle?*

A It is an area in which so many ships and planes have disappeared that some people call it the Devil's Sea. The area stretches from Cape Hatteras (North Carolina) 570 miles out to the island of Bermuda, and reaches south to Florida and the outer islands of the Caribbean.

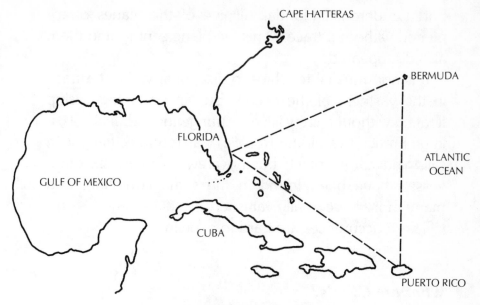

THE BERMUDA TRIANGLE

Q *Why are there so many accidents in the Bermuda Triangle?*

A Most scientists agree that it's because that particular section of ocean has an enormous amount of traffic all year, so it's natural that it would have more accidents.

However, many people feel that there must be some mysterious force at work in this area. In 1945, five U.S. Air Force planes set off on a routine training flight from

Fort Lauderdale, Florida. All five of the planes disappeared without a trace. A rescue plane sent to find them also disappeared.

It is not unusual for ships to disappear without a trace in the vastness of the ocean. But an undamaged ship floating without a crew is very strange indeed. In 1840, a large French vessel, the *Rosalie,* was found drifting, with sails and cargo intact, but no crew. At least six other vessels have been found in the same condition, and many more have totally vanished.

These occurences are hard to explain.

Q *What are UFO's?*

A The initials stand for *Unidentified Flying Objects,* and unidentified they remain. Thousands of people have reported seeing glowing objects in the sky. Some even say that the objects have come down to earth, and that space creatures have visited with them, and then taken off again. It is said that their vehicles sometimes leave scorch marks on the earth and a smell of sulfur in the air.

Q *Do UFO's really exist?*

A Nobody really knows. The United States government heard about so many "sightings" that it conducted a scientific investigation. The result? They could neither prove nor disprove that the mysterious lights in the sky were optical illusions caused by lens-shaped clouds or high-flying balloons or unexpected comets or meteorites. Perhaps UFO's are our space-age version of fairies and giants. Perhaps not.

Q *Is there magic in numbers?*

A Many people think so. Three, seven, nine, and thirteen are the numbers most often used in spells and legends.

Three is lucky. The third child in fairy stories is often the luckiest. A person often has a chance at three wishes. (We still say, "Third time lucky!" in games of chance.) In baseball, the batter has three strikes before he's out.

Some people believe so strongly that thirteen is unlucky, they leave out the number "13" on skyscraper elevators!

Just for fun, try this game with the number 9 (which is, of course, 3 times 3). Multiply any number by 9, then add up the total of the digits in the answer: They always come out to a number whose digits add up to 9. (For example:

$9 \times 565 = 5,085$. Five plus zero plus 8 plus 5 equals 18. Eighteen $(1 + 8) = 9$.

Nobody knows much about magic. But you can certainly sharpen your arithmetical skills by playing "magic" games with number 9!

ODD FACTS

From breaking the sound barrier to goats and sheep employed as living lawn mowers by the U.S. Army, from ancient Egyptian mummies to acid rain, from an iceberg the size of Belgium to dogs that make parachute jumps—the world is indeed full of a number of curious things!

Here are some odd facts to amuse and instruct you.

You'll find plenty more in books, magazines, and newspapers if you keep checking!

Q *Where does the word "dollar" come from?*

A From a place in Germany called Joachimsthal. Five hundred years ago the King of Bohemia (an area now on the borders of East Germany and Czechoslovakia)mined silver for his coins in Joachimsthal. At first the coins were called Joachimsthaler. Then the name was shortened to thaler or taler. Later, when other countries adopted the coins as money, they found it convenient to call the coins talers, sometimes pronounced dalers, then dolars, and finally, dollars.

Q *What is the sound barrier?*

A It is the increased resistance an aircraft meets when it travels at or near the speed of sound.

Q *What is the speed of sound?*

A About 74 m.p.h., but it varies with air temperature and altitude.

Q *Has anyone ever flown faster than the speed of sound?*

A Yes, many times. William Knight (project X-15, U.S.A.) flew his rocket at 4,520 miles an hour—that's 75.3 miles a minute! That's Rome to New York in less than an hour; and New York to Chicago in minutes.

Q *What is the fastest speed at which any human has traveled?*

A According to present records, it's 24,791 miles per hour, achieved by the command service module of *Apollo 10*, in May 1969, at an altitude of 400,000 feet above the ground.

Q *Has a winged aircraft ever been flown into space?*

A Yes. In April, 1981, the first space shuttle, *Columbia*, flew into space, piloted by two men, and returned to a safe earth landing. The U.S.–launched *Columbia* was the first aircraft to fly into space, return, and be used again on subsequent space missions. It is also the first spacecraft to be commercially operated. That is, people can buy space on the shuttle which may be used for experimental or other purposes.

Q *What is the world's fastest train?*

A France's TGV (short for *Train à Grande Vitesse,* or High Speed Train). This sleek, torpedo-shaped train, put into service in September, 1981, does an average of 156 miles (260 kilometers) per hour along special tracks. Unlike most trains, it can climb steep hills, so it doesn't need tunnels.

Q *How can the TGV move so fast and zoom up and down hills?*

A It is run by electricity, so it doesn't need to carry large amounts of fuel. It receives its power from its special track and overhead cables. Its motors are rated at about 8,500 horsepower.

Q *How do the holes get into Swiss cheese?*

A The holes are caused by bacteria that generate carbon dioxide gas. The gas forms "bubbles," as it does in carbonated beverages such as soda, around which the curd settles.

Q *How does a time-release capsule know when to release?*

A Each tiny particle of medication is enclosed in a layer of gelatin of a different thickness. The thinner the layer, the more quickly it dissolves in the stomach. The thickest layers may take up to twelve hours to dissolve and release the medication.

Q *What do tree trunks and fish scales have in common?*

A Both fish scales and tree trunks grow in rings. The older the fish, or tree, the more rings the scale, or tree trunk, will have. Of course, tree rings are easier to "read," since they are large. You have to look at most fish scales through a microscope to see how they have grown.

Q *How many people are alive in the entire world today?*

A According to some calculations, about 4.4 billion. People make these calculations by studying the census, or head counts, that are made by governments in most countries. But not everybody reports to the census bureau, and thousands in remote areas, such as the islands off New Guinea where a new tribe of people was discovered recently, have never even heard of a census. So the world population count can be only guessed at.

Q *How many people have ever lived in the world, since* **Homo** *sapiens (our species) first developed on earth about 3 million years ago?*

A That's a question almost impossible to answer, but an intelligent guess, made by a professor of ecology, is 50 billion people. That means that the 4.4 billion people alive today constitute about 9 percent of all the *Homo sapiens* who ever lived.

Q *What is meant by "the population explosion"?*

A It is a term used to describe the dramatic increase in the world's population in comparatively recent times, especially since the beginning of the twentieth century.

In the year 8000 B.C., the dawn of agriculture (as opposed to hunting and gathering), the estimated population of the world was about 5 million people.

In A.D. 1 it was about 200 million; this is a fairly accurate figure, since early scientists were already documenting and recording population growth.

In 1850, the beginning of the industrial age, there was a population of one billion. Less than one hundred years later, in 1945, the end of World War II and the beginning of the nuclear age, it had more than doubled to about 2.3 billion people. And today, a mere forty years later, it has almost doubled again to 4.4 billion.

Thus, a "growth curve" of the world's population would look something like this:

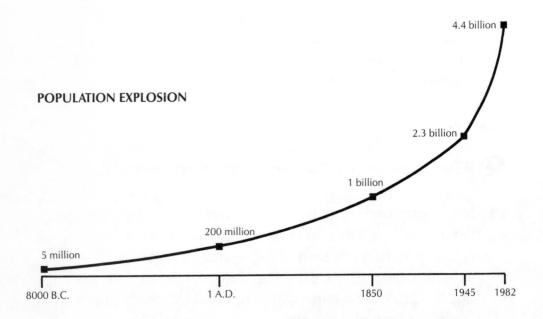

POPULATION EXPLOSION

Q *What is a census?*

A It's a head-count of all the people living in a country. The latest one in the United States was in 1980. It is said that if all the census forms sent out were laid end to end, they would encircle the globe three times!

Q *What is the Domesday Book?*

A It was a land survey and census ordered by William the Conqueror, the Norman invader who conquered England in the year 1066. He wanted to find out exactly how many people and how much property he had under his rule, so that he could figure out how much money he could take in taxes. The remarkable thing about the Domesday Book is that it still exists! It is a valuable historical record for those who want to know facts about life in eleventh century England. (The Domesday Book may be seen at the Museum of the Public Record Office in London, England.)

Q *Why was the eleventh-century head-count of England called the "Domesday Book"?*

A To frighten people into giving an honest report about their possessions. The word *domesday,* pronounced doomsday, means the Day of Judgment. In Biblical terms, the Day of Judgment occurs when everyone who has ever lived will be judged by the Lord. The fear of being caught in a lie encouraged people to tell the truth about their property, even though it meant paying higher taxes.

Q *How can an aerosol spray affect the earth's climate?*

A The *fluorocarbon* chemicals in any aerosol spray shoot up into the earth's stratosphere and destroy a part of the valuable layer of ozone that helps to protect the earth from the sun's dangerous ultraviolet rays.

Increases in the amount of ultraviolet light can help to cause skin cancer and damage to plant life and maybe even trigger climate changes.

(Refer to the chapter on Planet Earth for more about "the greenhouse effect.")

Q *How did a fish no longer than your little finger interrupt the building of a $130-million dam?*

A By being a very rare species: the snail darter, as the little fish is called, was thought to exist in only one stream in the world. That stream was about to be swept out of existence by the building of the Tellico Dam in Tennessee in the late 1970's. For two years environmentalists delayed the construction of the dam to save the fish. Then some more snail darters were found in a river 80 miles away. The threatened snail darters were transferred to the new habitat, and the dam was built.

Q *Where are you most likely to find living, breathing, warm-blooded lawn mowers?*

A At Fort Belvoir, a United States Army base in Virginia, and in other Army bases all over the country. The "lawn mowers" are goats and sheep that nibble on grass all day long, keeping it very trim. These animals are used because sparks from mechanical mowers might cause nearby ammunition to explode.

Q *What is the four-minute mile?*

A Four minutes was the record-breaking time that it took Englishman Roger Bannister to run one mile. The date was May 6, 1954. His actual time was 3 minutes, 59.4 seconds.

Q *What made the four-minute mile so historic?*

A It was the first time that anyone had run a mile in four minutes. Until that time, the record had stood at Sweden's Gunder Haegg's 4 minutes, 1.4 seconds, run in 1945. People thought that it was impossible for anyone to break that record.

Q *Why is it that once the four-minute "barrier" was broken, people began to run faster and faster?*

A One answer is undoubtedly that competitive people always want to break existing records. Once they know that something is possible, they are determined to do it themselves. Another answer is that as people learn more and more about the human body and how it works, they eat better and are more skillful about training their muscles.

Q *Is it possible that a three-minute mile may be run?*

A Yes. The latest world record for the mile is 3.47.30 minutes, run by Sebastian Coe (U.K.) in Brussels in August, 1981. Keep checking the records, and one day there may be a mile run in three minutes!

Q *Is 13 an unlucky number?*

A No. According to scientific reasoning, no number is either lucky or unlucky. But in many countries all over the world people are superstitious about the number 13. In some buildings the number 13 is omitted, and elevators seem to go magically from the 12th to the 14th floor!

But now take a look at the back of a $1 (one dollar) bill. There is a picture of a pyramid with 13 steps. Opposite is an eagle clutching 13 arrows and an olive branch with 13 leaves. The shield in front of the eagle has 13 stripes.

All of these thirteens symbolize the 13 original states.

There is nothing unlucky about having a dollar to spend—and having $13 is even better!

Q *Where did the idea of using coins as money originate?*

A Nobody is quite sure, but it was most likely somewhere in Africa. Money is something used to trade with. Person Number 1 wants something that Number 2 has. Let's say it's a rare seashell. Number 2 says, "Okay, I'll give you my seashell if you'll give me your matchbook cover from the Ritz Hotel." That's called trading. People all over the world used to exchange goods or services. Through the centuries, people got the idea that making coins out of precious metals such as gold and silver was a lot more convenient than carrying around huge stones or herds of cattle or even bags of gold dust.

Q *Where was the world's largest bathtub?*

A In ancient Rome. Using slave labor, the Romans built huge public baths. The largest were the Baths of Caracalla, built in the third century. They covered thirty-three acres of land. Sixteen hundred Romans could bathe at the same time. The ruins of the Baths may be seen in Rome to this day.

Q *Do men and women still bathe together in public baths?*

A Yes. In Japan, public bathing has been popular for centuries. It is considered an enjoyable and social way to relax at the end of the day.

Q *Where in the world do people check into coffinlike capsules rather than regular hotel or motel rooms?*

A In Tokyo, Japan. Capsulelike sleeping areas are becoming more and more popular as an economical, "no-frills" way for a businessman or tourist to spend the night away from home. The capsule contains a bunklike bed. There is just enough room to sit up without bumping your head, and room for your arms to stretch out to control switches for lights, service, and television. Dressing and undressing is done in a locker-room area.

Q *How did the jeep get its name?*

A The name derives from the initials G.P., short for general purpose. The vehicle was perfected during World War II. General George C. Marshall called it America's greatest contribution to modern warfare.

Q *What is an Egyptian mummy?*

A It is the long-dead body of a person (or sometimes an animal, such as a cat) that was preserved immediately after death and wrapped in cloth. Mummies were usually buried in painted boxes or even golden caskets, which were placed inside stone tombs.

Q *How were mummies preserved?*

A The methods of preserving the dead varied according to the period and to the wealth of the family.

In ancient Egypt the commonest method was embalming.

Often the intestines (called entrails) were removed and preserved in a separate jar. The cavities of the body were filled with resins and spices. Then the body was immersed for several weeks in a soda solution. Finally, it was wrapped in numerous strips of linen.

If the conditions were favorable (very dry, as in Upper Egypt) mummies unearthed and examined five thousand years later still had skin, hair, and bones. Most of the mummies found in humid Lower Egypt have disintegrated.

Q *Why did the ancient Egyptians and other peoples want to preserve their dead in such an elaborate way?*

A Because their religion led them to believe that at some future time the dead would come back to life, and would have need of their bodies, as well as food, weapons, wealth, sacred cats, and other animals. Of course, the richer the person, the more magnificent were the buried treasures and the larger the tombs. The great pyramids near Cairo in Egypt and the golden treasures of the boy-king Tutankhamen and others were made for very rich and powerful kings and queens.

Q *Why does the word "love" mean "zero" in tennis scores?*

A It may come from the French word, *l'oeuf,* meaning egg. In France, where an early form of tennis was invented, the word *l'oeuf* was a slang word for zero. So the winner of a point would say, "Fifteen for me and an egg for you."

When the game was adopted by the English, they said "Fifteen for me and *l'oeuf,* (or luff), to you." The pronunciation gradually changed to "love."

Q *Who wrote the song, "Happy Birthday To You"?*

A Two sisters, Mildred and Patty Hill, wrote a song called "Good Morning To You" in 1893. No one paid much attention to it until someone at a party changed the words to the ones we know today. Now it's sung all over the world all through the year.

Q *Is there any difference in taste between a white egg and a brown egg?*

A No. They are identical in nutrition and taste. The only difference is that they are laid by different breeds of hen. For example, Rhode Island Red hens lay brown eggs. Leghorn hens lay white eggs.

Q *What is acid rain?*

A Acid rain is rain that contains dangerous pollutants. The pollutants are caused by nitrogen and sulfer spewed out into the atmosphere in recent years by large industrial plants using oil and coal. The pollutants are dissolved into the atmosphere. Then, when it rains, the raindrops are full of sulfuric, nitric, and hydrochloric acids.

Q *Why is acid rain dangerous?*

A Because it poisons fish, frogs, salamanders, and probably other kinds of animal and plant life.

Q *How can sex be used to control insect pests?*

A Certain chemical substances are produced in an animal's body when it is ready to mate. These hormones send out powerful messages to attract a partner of the opposite sex. In many insect species these hormones, or pheromones, will attract male insects for miles around. Scientists have learned how to produce certain insect pheromones in the laboratory. The artifically produced pheromones lure insects into a trap, where they die.

Q *Why does hot water clean better than cold?*

A Because most materials dissolve more easily in higher temperatures. Also, heat speeds up the chemical reaction which binds soap to particles such as grease. In addition, hot water helps to kill most bacteria (tiny animals, invisible to the naked eye, that live on and in most substances). That is why surgical instruments are boiled in water before and after use, and why dishes must be washed in hot water.

Q *When is it better to rinse an article in cold water rather than in hot water?*

A When you have a bloodstain. Because of a chemical reaction, hot water tends to bind blood more firmly into a fabric.

Q *What makes bread become moldy?*

A Tiny mold spores (a kind of seed), usually present in the air all around us, tend to settle on bread (and other foods) that are left unrefrigerated, or even if kept in the refrigerator for a long time. They quickly develop into colonies that spread all over the food. Chemical substances added to the food (additives) help prevent the growth of mold spores.

Q *How big is the biggest iceberg ever seen?*

A It's hard to measure these floating giants accurately, so reports vary, but the biggest one ever spotted, near Antarctica, was in 1956. Some say that it was 208 miles long and 60 miles wide. This would make it 12,000 square miles in area; that is larger than the country of Belgium.

Q *What is the coldest place on earth?*

A The lowest air temperature ever recorded was at the Russian station Vostok in Antarctica, in August, 1960. The temperature was $-126.9°$ F ($-88.3°$ C).

Q *Which country's name never appears on its postage stamps?*

A Great Britain. Only the price of the stamp and a portrait of the sovereign appear on British stamps. This practice dates back to the first adhesive postage stamps, the British Penny Black and Twopenny Blue, advocated in 1840 by Sir Rowland Hill. They bore the image of the reigning monarch, Queen Victoria. Since then, Britain has always considered a likeness of the head of its sovereign to be sufficient identification for the country of origin.

Q *Have all the animals and plants on earth been discovered and named?*

A No, far from it. Every year scientists and explorers discover new species of birds, mammals, insects, fish, and even hitherto unknown tribes of human beings. There are still many remote islands, valleys, hilltops, and forests where few people go. Natives who hunt there may know the creatures and plants that live there, and have local names for them. But it takes an expert scientist to study and identify a "new" animal or plant and classify its relationship to known fauna and flora.

Q *What event was described in these words: "NEW YORKERS HAVEN'T SEEN ANYTHING LIKE THIS IN 9,500 YEARS"?*

A The birth of Astor, an elephant born at the Bronx Zoo, New York, in the summer of 1981. The son of Groucho and Patty, Astor weighed 200 pounds at birth. He started to walk after only 25 minutes and appeared shyly on television a few days later, carefully guarded by his protective mother.

The 9,500 years referred to in the headline was presumably the last (estimated) time that a Siberian or a woolly mammoth, the modern elephant's ancestor, was born in the New York geographical area.

Q Is the U.S. Army training dogs to jump by parachute?

A Yes. A special harness has been developed that allows a dog to be attached to its handler just beneath the reserve parachute. In a jump, the dog rests against the handler's legs in its harness, leaving the soldier's hands free. At a distance of about 200 feet (60 meters) from the ground, the handler lowers the dog on an 18-foot (6-meter) line.

Army dogs (mostly German shepherds) are useful in many ways, including tracking people, bombs, and narcotics.

Q What is a carat?

A It is a universal measurement for weighing gold and gems, and comes from the Greek word for carob tree, whose uniformly sized seeds, or beans, were used to balance scales in ancient bazaars. In the international system of weights and measures, 142 carats equal one ounce. Pure gold is 24 carats. Most American jewelry is 14 carats, which means that it is 14 parts pure gold, and 10 parts alloy—another metal, such as copper or silver. Carat is often spelled with a k: karat.

Q *What was the purest, largest nugget of gold ever found?*

A The "Welcome Stranger," weighing 2,280 ounces (142½ pounds). It yielded 2,248 ounces of pure gold from a total weight of 2,280¼ ounces. It was found in Moliagul, Victoria, Australia.

Q *What part of a cat is used to make the catgut strings on musical instruments?*

A No part of a cat. Catgut is made from the dried intestines of sheep, pigs, and other animals. Nobody knows how the word "cat" got into the act! Catgut is very strong. It is used also for making the strings on tennis rackets and for surgical sutures (stitches). The catgut slowly dissolves into the body, so the stitches don't have to be taken out. Today, new man-made materials are taking the place of many catgut products.

Q *Can works of art be preserved by nuclear radiation?*

A Yes. Nuclear physicist Ioan Crihan discovered that tapestries, oil paintings, and wood sculptures are teeming with destructive bacteria and termites. Crihan calculated that an exposure to radiation even 5 million times as powerful as an ordinary chest X-ray would not harm wood, canvas, or tapestry, but would kill the harmful animals. Early in 1981 he patented the first art-preservation technique based on nuclear radiation.

Q *How old do fish get?*

A Nobody knows for sure, because there are so many millions of fish that nobody can keep tabs on them. There are, however, some accounts from people who have kept and observed fish in aquariums and backyard ponds. Here are some examples: catfish: 60 years old; carp: 50 years old; halibut: 30 years old; trout: 18 years old.

BOOKS AND MAGAZINES TO LOOK UP

Magazines:
AUDUBON SOCIETY MAGAZINE
NATIONAL GEOGRAPHIC MAGAZINE
NATURAL HISTORY MAGAZINE
SCIENCE NEWS
SCIENTIFIC AMERICAN
SCIENCE 81 and SCIENCE 82

For Younger Readers:
RANGER RICK'S NATURE MAGAZINE
NATIONAL GEOGRAPHIC WORLD MAGAZINE

FURTHER READING

Books:

THE AUDUBON SOCIETY ENCYCLOPEDIA OF NORTH AMERICAN BIRDS, Random House, New York, 1980

THE BIOLOGY OF FLOWERS by Eigil Holm, Penguin Books, London, 1979

BODY WORDS by Kathleen N. Daly, Doubleday, New York, 1980

THE ENCYCLOPEDIA OF HOW IT WORKS, edited by Donald Clarke, A&W, New York, 1977

THE ENCYCLOPEDIA OF PREHISTORIC LIFE, edited by R. Steel and A. Harvey, McGraw-Hill, New York, 1979

ENCYCLOPEDIA OF THE ANIMAL KINGDOM, R. & M. Burton, A&W (Galahad) Books, New York, 1976

FLOWERS OF THE WORLD by Frances Perry, Crown, New York, 1972

THE GUINNESS BOOK OF WORLD RECORDS, edited by Norris McWhirter, Bantam Books, New York, 1980

LIFE ON EARTH by David Attenborough, Little Brown and Company, Boston, 1981

THE MACMILLAN YEAR BOOK 1981, Macmillan Educational, New York, 1980

THE NEW COLUMBIA ENCYCLOPEDIA, Columbia University Press, New York, 1975

1001 QUESTIONS ANSWERED ABOUT BIRDS by Allan and Helen Cruikshank, Dover Books, New York, 1976

1001 QUESTIONS ANSWERED ABOUT EARTHQUAKES, AVALANCHES, FLOODS, AND OTHER NATURAL DISASTERS by Barbara Tufty, Dover Books, New York, 1969

FURTHER READING

THE POCKET GUIDE TO ASTRONOMY by Patrick Moore, Simon & Schuster, New York, 1980

THE RANDOM HOUSE ENCYCLOPEDIA, Random House, New York, 1977

STORIES BEHIND EVERYDAY THINGS, The Reader's Digest Association, Inc., New York, 1980

THE SUPERNATURAL, The Danbury Press, Grolier, and Aldus Books, London, 1976

THE TIME-LIFE NATURE LIBRARY, Time-Life Books, Inc., Alexandria, Va.

THE TIME-LIFE SCIENCE LIBRARY, Time-Life Books, Inc., Alexandria, Va.

INDEX

The figures in *italics* refer to illustrations.